NARRATIVES *of* EDUCATING *for* SUSTAINABILITY *in* UNSUSTAINABLE ENVIRONMENTS

NARRATIVES *of* EDUCATING *for* SUSTAINABILITY *in* UNSUSTAINABLE ENVIRONMENTS

EDITED BY
Jane Haladay AND Scott Hicks

Michigan State University Press | *East Lansing*

♾ The paper used in this publication meets the minimum requirements
of ANSI/NISO Z39.48-1992 (R 1997) (Permanence of Paper).

Michigan State University Press
East Lansing, Michigan 48823-5245

Printed and bound in the United States of America.

26 25 24 23 22 21 20 19 18 1 2 3 4 5 6 7 8 9 10

LIBRARY OF CONGRESS CATALOGING-IN-PUBLICATION DATA
Names: Haladay, Jane, editor. | Hicks, Scott (Brian Scott), editor.
Title: Narratives of educating for sustainability in unsustainable environments /
edited by Jane Haladay and Scott Hicks.
Description: East Lansing : Michigan State University Press, 2017.
| Includes bibliographical references.
Identifiers: LCCN 2016057875| ISBN 9781611862645 (pbk. : alk. paper) | ISBN 9781609175467 (pdf)
| ISBN 9781628953152 (epub) | ISBN 9781628963151 (kindle)
Subjects: LCSH: Sustainability—Study and teaching (Higher) | Environmentalism—
Study and teaching (Higher) | Education, Higher—Environmental aspects.
Classification: LCC GE196 .N37 2017 | DDC 304.2071/1—dc23
LC record available at https://lccn.loc.gov/2016057875

Book design by Charlie Sharp, Sharp Des!gns, East Lansing, MI
Cover design by Shaun Allshouse, www.shaunallshouse.com
Cover image, "Boat on Forstall Street," Lower 9th Ward of New Orleans,
©2007, by John Rosenthal and is used courtesy of the photographer
(www.johnrosenthal.com). All rights reserved.

Michigan State University Press is a member of the Green Press Initiative and is committed to developing and encouraging ecologically responsible publishing practices. For more information about the Green Press Initiative and the use of recycled paper in book publishing, please visit www.greenpressinitiative.org.

Visit Michigan State University Press at *www.msupress.org*

. . .

To university students
and their educators everywhere:

May we all work together
to sustain each other and earth
in the places we inhabit
and that inhabit us.

. . .

Contents

PART 2. Rethinking What We Do, Remaking Curricular Ecologies

PART 3. Reinhabiting and Restoring Who and Where We Are

Acknowledgments

We are grateful to the contributors in this collection for sharing the stories of their experiences in higher education, experiences that reflect not only the best of academe (the spaces of innovation, the partnerships that blossom from collaboration, the protections of tenure that foster critical scholarship) but the worst as well (the constrictive contingencies of adjunct labor, the displacement of faculty that hinders deep engagement with people and places, the administrative pressures to bend teaching and research in the pursuit of fundraising). These contributors' courage to speak personally and honestly of their experiences, for better and for worse—and sometimes critically of the institutions for which they work—has been the purpose of this project and has commanded our admiration from the start.

We thank Julie Loehr, assistant director and editor in chief of Michigan State University Press, for her unflagging support, guidance, and assistance in this undertaking; the encouraging remarks of our anonymous peer reviewers; and MSUP's team, who took this work from a computer file to a book.

In authoring our chapter on southeastern North Carolina, we thank Chad Locklear, author of "Swamp Posse," for sharing with us the inspiration and consciousness that informed his short story, a piece that has engaged us and our students and focuses us on where we are and how we might imagine new futures. We hope he continues to write fiction about southeastern North Carolina that we may bring to our future students.

We thank photographer John Rosenthal of Chapel Hill, North Carolina, for granting permission to use his compelling photograph, "Boat on Forstall Street," for the cover of our collection.

For both material support and personal encouragement, we thank the chair, Mary Ann Jacobs, and administrative support associate, Alesia Cummings, of the Department of American Indian Studies at the University of North Carolina, Pembroke. We also thank Rose Stremlau, assistant professor of history at Davidson College, formerly of UNCP, for her ongoing support and ideas about this project. To our many personal friends and professional colleagues who shared their enthusiasm for this collection as it came into being, we extend our heartfelt thanks.

We thank the Teaching & Learning Center of the Office of Academic Affairs at UNCP for generous grants that supported the writing, editing, and preparation of this book. As well, we are grateful to the University's Esther G. Maynor Honors College, under the leadership of Dean Mark Milewicz, for giving us the opportunity to team-teach environmental literature and thus to maintain the cooperation and thinking that undergird the ideas and hopes contained in this book.

And, especially, we would like to acknowledge one another, as colleagues, co-editors, co-instructors, and good friends, for bringing into reality the idea of assembling a collection on higher education and sustainability. Sharing each other's humor, intellectual perspectives, many cups of coffee, and a range of healthy and unhealthy snacks while working on this project over the years has made creating this collection a genuine delight—and set a high bar for future collaborations.

It is our hope that this book reminds you, our reader, that others are experiencing the opportunities and challenges that confront you as sustainability educators and advocates; that such consilience might reinforce you as you teach, write, and serve on behalf of the planet and all who inhabit it; and that together we can work within and despite the institutions that employ us and the politics—local, national, and global—that surround us to make the differences that matter.

Preface

Against its horizon,
I spread my arms like a road sign
To mark earth where we are.

—James Applewhite, "Observing the Sun"

Sustainability has become "one of those words," equally at home in a McDonald's ad as in high-level talks on climate change at the United Nations, under whose auspices it emerged in 1987. As the focus of the World Commission on Environment and Development's *Our Common Future*, sustainability entails "humanity['s] ability . . . to ensure that it meets the needs of the present without compromising the ability of future generations to meet their own needs." It has come to embrace anything and everything, inside and outside the academy alike, about which assumptions and presumptions are made depending on reader, speaker, audience, and context. The concept, often articulated as a "triple bottom line" of people, planet, and profits, or ecology, economy, and equity, transcends disciplinary bounds, signifying in science, philosophy, economics, and other fields of study, both theoretical and pragmatic. "Sustainability" is at once institutional and personal, systemic yet individual, as much a new (or old, depending on one's perspective) way of doing business that leaves something for the next generation as it is a personal philosophy to persevere. Depending, again, on one's perspective, it is either radical—a vision for the future of the planet that will turn the tide of ecological collapse, requiring mass action and revolutionary change—or tepidly conservative, a flaccid acquiescence to the catastrophes already underway, sustaining the devastation already wrought. From an Indigenous perspective, "sustainability" is a simplified, catchall term for an ancient and integrated system of sophisticated philosophies and practices that long ensured ecological vitality in place prior to colonization's ruptures. In the lived experiences of citizens of Indigenous nations, sustainability is nothing new. Instead, "the realization that the total community must be engaged in order to attain sustainability comes as a result of surviving together for thousands of years," Okanagan scholar and activist Jeannette C. Armstrong explains. "The practical aspects of willing teamwork within a whole-community system clearly emerged from having to cooperate in order to survive."[1] In short, "sustainability" is, metaphorically speaking, a slippery fish sometimes darting vigorously in a viable ecosystem, sometimes convulsively gasping in an oil-slicked tide pool.

Within higher education, the term can mean different things to different individuals, departments, and units—complicating who, how, and where sustainability is taught and studied. As Stuart Pearson, Steven Honeywood, and Mitch O'Toole note, "The contested nature of the language [of sustainability and environmental education] occurs within a situation that is already complicated by attempts to capture the field from various defended areas of scholastic turf." For scholars in ecology and environmental studies, sustainability refers to "the use of an organism, ecosystem, or other renewable resource at a rate that is within its capacity for renewal," part of a system of environmental relationship that seeks "the maintenance or restoration of the composition, structure, and processes of ecosystems" with the goal of "the benefit of future generations." For scholars working specifically in the field of sustainability, it might mean the consideration of natural resources in "enabl[ing] people and communities to provide for their social, economic, and cultural well-being and for their health and safety." From its scientific and technical origins, "sustainability" has taken root in business, economics, and public policymaking in the form of promoting "development that takes full account of the environmental consequences of economic activity and is based on the use of resources that can be replaced or renewed and therefore are not depleted." Scientific interests in the continuance of life, coupled with economic and political interests in determinations of balance, make sustainability a ripe field for philosophers as well, for whom it becomes a moral problem with an ethical imperative:

> If "development" means economic growth, this can bring benefits for some and disbenefits for others (e.g., unemployment and displacement due to new forms of industrialization). Development ethics recognizes that policy-makers, aid donors, corporations, and agencies . . . confront moral questions when planning socio-economic changes, particularly in the world's poorest countries.

No matter the field or discipline, however, "sustainability adds an important time dimension," for "changing human actions and interactions into the future permeates every prescription to promote movement toward sustainability."[2]

In some ways, what sustainability is not—and why it is urgent today, in the so-called Anthropocene—is put succinctly and immediately by Jeremy L. Caradonna, in his comprehensive overview, *Sustainability: A History* (2014). Caradonna writes that sustainability is not

> a moribund economic system that has drained the world of many of its finite resources, including fresh water and crude oil, generated a meltdown in global financial systems, exacerbated social inequality in many parts of the world, and driven human civilization to the brink of catastrophe by unwisely advocating for economic growth at the expense of resources and essential ecosystem services.

United across disciplines, approaches, and practices, "sustainists"—Caradonna's label for those from "scientists and engineers to economists, educators, policymakers, and social activists" dedicated to meeting the challenge of sustainability—are joined, similar to the Okanagan concept of whole-community cooperative systems, by their commitment to the lives and well-being of future generations of human and nonhuman life alike; to green, resilient economies; and to democracies that uphold social justice and peace. Human beings stand at a crossroads, Caradonna asserts, with "two paths to take: continue with business as usual, ignore the science of climate change, and pretend that our economic system isn't on life support or remake and redefine our society along the lines of sustainability." No matter its disciplinary shadings, a shared sense of purpose anchored in a repudiation of what has gone wrong drives the sustainability movement forward.[3]

The concept "sustainability," like this collection, thus embraces technical, material, as well as abstract definitions—the personal, emotional, and psychological as well as social, economic, and political. "Sustainability is also concerned with intragenerational equity: understanding the links between poverty and ecosystem decline," Derek Owens asserts. Like Caradonna, Owens states that "Sustainability means recognizing the short- and long-term environmental, social, psychological, and economic impact of our conspicuous consumption. It means seeking to make conservation and preservation inevitable effects of our daily lives." As academics, we need to view the institutions of higher education at which each of us works as "hav[ing] a responsibility to address a nexus of systemic economic, environmental and social problems that imperil human and non-human futures by placing 'education for sustainability' (EfS) at the centre of their concerns," as Fred Gale et al. contend. The authors rightly argue, "This responsibility requires not just diagnosis and treatment of problems outside of the academy, but also critical analysis on the question of how higher education is implicated in the making of these problems." To put a finer point on it, "The people who are now leading us down an unsustainable path have come out of the best universities in the world," Margaret Robertson writes. "These institutions have armed us with knowledge suited for the conquest of nature, with the result that we are now misfits on our home planet. Yet higher education holds opportunity [and] colleges and universities have tremendous leverage."[4] As we generate these analyses, engage in meaningful scholarship, and craft dynamic curricula, how do we—as vibrant academics and healthy human beings—also carve out sustainable and personally gratifying lives within limiting institutions and challenging contexts?

The personal and professional stories of the authors in this collection challenge the status quo of higher education and offer a collective response to these questions. In each narrative, the authors of the essays that follow ask, How do we educate for sustainability, and what does it even mean to do so from the primarily humanistic perspective reflected in these chapters? How—which the editors consider equally and possibly even more importantly here—do we sustain ourselves as educators in the systems, places, and ecologies that we find ourselves teaching in as adjunct and tenured academics? How can, and how do, many of us go on in these systems—systems that external groups have targeted variously for transformation, reform, or elimination?

Through pedagogical narratives, literary analyses, reflective essays, and collaborative dialogues, *Narratives of Educating for Sustainability in Unsustainable*

Environments explores the professional and intellectual tensions of curricula, pedagogy, and personal practice that honor the relationships of interspecies ecologies, reinhabit and reconceive wounded landscapes and wounding institutions, and help us to reattune ourselves to new yet ancient frameworks for sustainability. Whether they live, teach, and write in Australia or Toronto, in Ames, Iowa, or St. Petersburg, Florida, the authors in this collection are united in the recognition that durable and resilient institutional, personal, and professional frameworks are critical to sustainability. Our collection confronts head-on the contexts that make environmental pedagogies difficult, the challenges to the well-being of the teacher-scholar, and the corrosive academic structures that compartmentalize knowledge and people. As it confronts them, our collection simultaneously offers models for working through and within these challenges in order to advance understandings and ways of being on local, global, and personal levels that can turn the planetary tide toward effective, shared, resilient sustainability. To bring about the "sustainable academy," Richard M. Carp calls for "transform[ing] the academy" by "find[ing] the assets for knowledge production available in every community in which we take place and . . . connect[ing] them in diverse and local ways to multiply their power and effectiveness." For Pearson et al., "The way forward here, as in so many other situations and locations, is through better and deeper connections between the people charged with doing this important work."[5] We offer this collection as one way of making those "better and deeper connections."

Jennifer Case opens Part 1, "Confronting the Challenges of the Places We Are," by discussing the experience of being a graduate assistant teaching place-based composition courses at a campus that was not a place she was from nor a place she would remain. "Unrooted: Dislocation and the Teaching of Place" describes the ways in which Case's time as a graduate student at the University of Nebraska-Lincoln led her to confront how much she is "affected by place" and to recognize that many undergraduate and graduate students, as well as faculty who relocate for their careers, are "unrooted" by their academic experience. Their displacement requires them not only to adapt to a new job or degree program but to new ecological, social, geopolitical, and cultural contexts, as Case explains. Higher education's "emphasis on the cognitive has worked to harm local regions—not only by overlooking their significance, but also by physically luring students away," Case argues, because it creates knowledge and skills that are meant to be geographically mobile. Graduate students, adjuncts, and temporary faculty who care about sustainability education are simultaneously in tenuous positions that impact their ability to dedicate themselves fully to teaching meaningfully about, and to personally connecting with, the places in which they find themselves for short or uncertain lengths of time. For Case, assigning her students a place-based personal narrative proved to be one way she could simultaneously create a curriculum that privileged specific places—those known, valued, and vividly described by students in their essays—and allow her to connect more meaningfully to the Nebraska landscape and to a range of homeplaces familiar to her students. While Case concludes that "we cannot expect our faculty members to be able to teach sustainability if they, themselves, are balanced precariously in unsustainable environments," she is clear that this is nevertheless what many of us are consistently asked to do.

In "By the Lumbee River with Chad Locklear's 'Swamp Posse,'" the editors of this collection describe the opportunities and challenges of beginning an environmental literature course with a story that tells of the place they are. On the one hand, the Lumbee author's short story celebrates the nature of the region, a blackwater river and floodplain in North Carolina's coastal plain. On the other hand, the story broaches the realities of racism, sexism, poverty, homophobia, and violence—persistent, entrenched, and ongoing challenges with roots in the historical trauma of colonialism and segregation of the region's racially and ethnically diverse population. When the story's protagonist, Vince, returns home to Robeson County from graduate school, his reunion with old friends for a night of drinking and talking by the Lumbee River reveals deep-seated tensions and anxieties that inform his ambivalence toward coming home. With Greg Garrard's comprehensive and forthright *Ecocriticism* (2012), we guide our students in thinking about this story ecocritically; as they continue to explore the story, we guide them to consider questions that drive our discussion of subsequent texts—questions that also lead them to consider more deeply the places they themselves are from as well as the place where they are now attending college. Finally, we discuss how teaching "Swamp Posse" speaks to our personal and professional struggle as faculty who have witnessed the best and worst of Robeson County during our ten years of teaching here, wrestling with our responsibility to the landscape in which we teach and the human community whose homelands we are in, our capacity to persevere as educators and citizens, and the lessons our students and their communities teach us as to how best we might all live here and share a common ground.

Daniel Spoth of Eckerd College on Florida's Gulf Coast writes, at seven feet above sea level, from the perspective of an English professor whose campus will be underwater by century's end. In "Getting Your Feet Wet: Teaching Climate

Change at Ground Zero," Spoth describes the urgency and challenge of a pedagogy that creates graduates who will become the leaders who confront climate change and take down the barriers that keep us from taking action. Whereas Eckerd and universities across the planet are finding ways to become more sustainable and resilient, even as political leaders like Florida governor Rick Scott deny climate change, Spoth challenges faculty to bring sustainability into the classroom for the transformative potential of environmental education. To inculcate a sense of place in his literature students, Spoth draws on writers and thinkers like Eudora Welty, Anthony Giddens, Bruce Robbins, and Yi-Fu Tuan and explores models of place-based attachment and the challenges of attachment and connection to many places toward what David W. Orr has called "the arts of inhabitation." The challenges are difficult: college students might come from thousands of miles away, graduate and leave for jobs another thousand miles away, or view higher education as nothing more than a steppingstone to increased consumption. And, Spoth notes, it is not enough to develop solely the individual. Rather, he asserts that our pedagogy must help students make community and global relationships that underscore the connections that constitute sustainability. He concludes that educating for sustainability must evoke, analyze, and transform students' emotions in the face of the paralysis that global climate change otherwise effects.

In "Cutting through the Smog: Teaching Mountaintop Removal at a University Powered by Coal," Brianna Burke narrates compellingly just how interconnected our energy systems and our personal and professional lives can be. A professor of environmental humanities at Iowa State University, Burke's job as faculty and the educational experiences of her students are energized by the ten-story coal power plant that fuels the school day and night, surging any time students, faculty, or staff flip on a light switch. Part of Burke's response to this awareness has been to create a course that strives to make coal energy—which she reminds us is "the largest contributor to global greenhouse gas emissions"—more visible to her students through an interdisciplinary curriculum that melds the voices of coal community members whose lives are economically, socially, and physically impacted by working in a coal economy with representations by coal corporations of the value and "clean" energy of coal production. Emphasizing the destructive force of mountaintop removal through a range of texts including the documentary *The Last Mountain*, Anne Pancake's novel *Strange as This Weather Has Been*, and the music of 2/3 Goat, Burke and her students explore the "insider" and "outsider" perspectives that argue all sides of coal extraction as a source of sustainable energy alongside scientific data and personal narratives (fictional and nonfictional) that humanize the stories of coal-mining communities. With its themes similar to those of Locklear's "Swamp Posse," Burke's course leaves students pondering the complexities of whether to stay or leave one's home community, wrestling with the question of responsibility for repairing the destructive system of coal extraction. Taking her students to tour Iowa State's power plant is the culminating activity of Burke's course, one she has seen make a lasting impact on students' thinking about the various stakeholders in coal energy and how we might work toward cleaner energy while honoring all constituencies.

Jennifer Schell's personal pedagogical essay, "Teaching about Biodiversity and Extinction in a Thawing Alaska: A Reflection," outlines the environmental, economic, and geopolitical obstacles she faces as a member of the English faculty at the University of Alaska Fairbanks (UAF). Schell describes the tensions of having "experienced, firsthand, the unsustainability of Alaska's economic dependence on fossil fuel industries," which fund nearly 90 percent of UAF's operational budget; Alaska, she notes, is the only state in the United States that does not collect sales or income tax to help support public education. Because of a sharp decrease in oil prices in recent years, Alaska finds itself facing a crisis regarding funding public services, including higher education. For those working at and attending UAF, Schell tells us, revenue shortfalls have meant "layoffs, furloughs, workload changes, and other budget-cutting measures," the elimination of courses and degree programs, a reduction in faculty travel budgets and an increase in faculty course loads (with less time for research and creative activities), and student athletes traveling to fewer events. Yet even as Alaska's ecological biodiversity and public services are radically eroded by the effects of fossil fuel extraction and human dependence on the oil industry, Schell has noticed an unwillingness on the part of some of her Alaskan students to critique this system and to take seriously the environmental threat posed by climate change, despite the range of texts she assigns in her "critical place-based" course focused on Alaskan biodiversity and extinction. By detailing her experience teaching one particular course, Schell outlines a larger and troubling landscape for universities dependent on corporate funding and for regions whose political leadership remains committed to economies based on environmentally unsustainable practices. At the time of this writing, Schell sees "no relief in sight" for the unsustainable working environment she faces at UAF.

The essays in Part 2 explore the theme of "Rethinking What We Do, Remaking Curricular Ecologies." The opening chapter is a moving and powerful collage by

Derek Owens, who explores the question of meaning-making in a world of irreversible ecological catastrophe in his essay "Letting the Sheets of Memory Blow on the Line: Phantom Limbs, World-Ends, and the Unremembered." What rituals and procedures can we use, Owens asks, to mourn mass die-offs and extinctions taking place on an incomprehensible scale, an unimaginable scale? Perhaps, he suggests, we are experiencing pre-traumatic stress disorder, a pervasive undercurrent of anxiety, dis-ease, and melancholy, the result of attempting to anticipate a real end times. Throughout his chapter, Owens draws on a rich range of sources—experimental memoir, trauma theory, poetry, environmental studies, philosophy, ecocriticism, and others—to imagine and reimagine memory, mourning, and making sense of ecological crisis. A former college administrator who has now returned to the classroom, Owens sees the power of pedagogy as the optimal site for reimagination and revision, in particular on a campus like his own St. John's University in Queens, New York, which boasts one of the United States' most diverse student bodies and is located at a global crossroads of ethnic, racial, linguistic, and spiritual human diversity. As Owens underscores, teaching and learning that embrace diversity of every sort, bringing into our classrooms multigenre, multimodal, multimedia experimental risk-taking, are the only rejoinder to the crushing hegemony of the way things are, the only response to ecological chaos and loss, and the only root of sustainability in an age of disappearing species, cultures, and economies. "In a time of hauntings and ghosts and phantom limbs—of anxieties born from the awareness that we are causing more extinctions than at any time since the Pleistocene—memory, trauma, violence all take on radically new auras," Owens writes. He concludes that how we teach, how we embrace our students, and how we tell new stories is our defining struggle in sustainability.

Since 2010, faculty from across Rose-Hulman Institute of Technology in Terre Haute, Indiana, have partnered in an interdisciplinary living-learning community called Home for Environmentally Responsible Engineering (HERE). In "Student Expectations, Disciplinary Boundaries, and Competing Narratives in a First-Year Sustainability Cohort," English professors Corey Taylor, Richard House, and Mark Minster describe the history, challenges, and future of this program, which seeks to give its graduates the tools and mindsets to become environmentally conscious engineers, and graduated its first cohort in spring 2015. As Taylor, House, and Minster have realized, it is difficult to lead a long-term, interdisciplinary program without course release time or an adequate institutional budget. Yet even more challenging might be the responses of students, who come to Rose-Hulman for its nearly 100 percent postgraduation employment rate and high starting salary, many of whom perceive learning about sustainability from non-engineering faculty as an irrelevant waste of time. In response, HERE's faculty have revised the program each year, attentive to students' responses and research in the field of engineering education while remaining true to the vision of HERE's core of environmental responsibility. At the time of writing their chapter, the authors note that HERE now foregrounds problem-based learning, and its sponsors have borrowed the market-speak of the college's brand that learning and doing sustainability sparks innovation and adds value for future engineering professionals. As the authors think about the future, they see the clear value of study in English, composition, and rhetoric to the work of sustainability—but they see even more clearly the problems hardened disciplinary silos present to durable progress in teaching, learning, and research in sustainability. Moreover, HERE faculty see that for too many, inside the academy and out, the value of education in sustainability continues to matter far less than making money. Taylor, House, and Minster conclude that sustainability programs in higher education must craft and share a compelling narrative about higher education and sustainability, one that shows the value of new ways of doing things to counter and recalibrate narratives focused on expectations of return on investment. By reclaiming higher education's myriad goals from the myopic narrative that equates money with success at the expense of everything else, the authors conclude that sustainability initiatives might be more effective and transformational in seemingly hostile (or indifferent) environments.

Andrea Olive reveals her hopes that educating for sustainability in her curriculum leads students to shape federal and provincial environmental policy in "Connecting Urban Students to Conservation through Recovery Plans for Endangered Species." Having spent much of her early life in the natural open spaces of rural Saskatchewan, Canada, Olive later found herself as a student and a faculty member at large urban universities in Calgary; Halifax, Nova Scotia; Washington, DC; Detroit; and now Toronto, where she holds appointments in political science and geography and planning at the University of Toronto. Olive recognizes that for many of her cosmopolitan students, biodiversity and urban development's impact on species extinction are invisible, able to remain so within the congested throng of city life. Olive's chapter details how she has found, nevertheless, in "this urban environment . . . the right place continually to find the inspiration to study endangered species conservation and sustainable living" and to make these issues palpable to students through analyzing biological data and conservation policies in Canada, where approximately 80 percent of the population is urban. Through creating "a recovery strategy for a federally listed endangered species that does not presently have a plan in place by Environment and Climate Change Canada"—species such as the Blue-grey Taildropper, Seaside Bone Lichen, and Bert's Predaceous Diving Beetle—Olive's students engage in detailed research that results in compiling existing data on their species, evaluating that data (or the lack of it, for many species), and making policy recommendations in their final reports. Over the course of this project, many students become fierce advocates of their species, and Olive has shared several of their recovery strategies with the Committee on the Status of Endangered Wildlife in Canada (COSEWIC), a group organized by the national government "to maintain a list of species at risk and [to be] responsible for writing recovery plans and action strategies for each endangered and threatened species—even those not found on federal lands." Olive's chapter describes how her environmental pedagogy, by encouraging each of her students to become passionate about a species they come to know well, strives to lead students to continue to think more expansively about the necessity and value of all species in an ecologically vital and sustainable world. Teaching about species extinction and recovery in a region of six million people is also, Olive tells us, how she is able to sustain herself personally and professionally between her returns during holidays to the open prairies she loves.

The humanities have a critical role to play in educating for sustainability, Keely Byars-Nichols of the University of Mount Olive writes in "Teaching Critical Food Studies in Rural North Carolina." The humanities, she claims, make us attentive to, and critical of, the paradoxes of hunger and poverty in a region of agricultural productivity, and the humanities enrich the experiences of diverse students, in particular at her campus, where many major in agriculture and related fields. In creating her critical food studies curriculum, Byars-Nichols introduces students to leading writers and thinkers such as Wendell Berry, Barbara Kingsolver, and Michael Pollan; guides them in writing reflective auto-ethnographies; and connects them with community partners for the purposes of alleviating hunger, connecting farmers to markets, and growing local economies for resilience. Throughout her teaching, she draws on local resources, like Kinston, North Carolina's Vivian Howard of *A Chef's Life* celebrity. Her students come to profound understandings of themselves, their families, and their communities through their auto-ethnographies of food histories and rituals, and they strive to make a significant contribution to their university's regional community, such as creating a campus and community sustainable organic garden with a local nonprofit, and fundraising for a farm-worker rights advocacy group. Byars-Nichols's chapter privileges the voices of her students as she outlines her curriculum, illustrating one way critical food studies pedagogy might restore to local foodscapes—especially those historically undervalued—the knowledges and histories of the people who work, grow food, eat, and live within them.

Part 3 of this collection, "Reinhabiting and Restoring Who and Where We Are," begins with Jesse Curran's reflection on the necessity of claiming and foregrounding sources of personal strength in developing professional sustainability. "Mindfulness, Sustainability, and the Power of Personal Practice" outlines her commitment to yoga and meditation, not only as practices that ground her own personal life, but as models for cultivating exactly the sort of mindfulness we ask of our students in sustainability higher education. Against the advice of a well-meaning faculty mentor, Curran has retained the term "practitioner of yoga" in her CV as an adjunct faculty member on the job market, because she believes that acknowledging this part of her identity asserts "one of the sources of strength" that allows her to sustain herself and flourish as an academic. Curran finds that "the practice of yoga and meditation has offered [her], to echo Thich Nhat Hanh, more concrete ways of making a commitment to sustainable living in [her] daily life," and she hopes to share these ways of being with her students through integrating yogic philosophies into her English and sustainability curriculum. The quality of having a "beginner's mind"—one that is open, questioning, and does not claim expertise or ultimate authority—is fundamental to yogic practice, and Curran discusses how this concept helps students embrace inquiry and challenge their own long-held preconceptions and behaviors in order to cultivate more sustainable practices. "The admission of uncertainty—of not having a full grasp of a given subject or situation—is understood as a position of intellectual and emotional freedom, as it suggests capaciousness rather than close-mindedness," Curran writes. "The large-scale imaginings of ecological enmeshment often ask their learners to dwell in uncertainty, and the beginner's mind allows students to entertain the unstable terrain of sustainability with flexibility and receptivity." Through integrating yogic and meditative philosophies in her curricula, Curran offers her students a meaningful approach to environmental education while also working to restore a more sustainable higher-education environment to counteract one that is consistently alienating, impersonal, and highly competitive.

"Ecological Journeys: From Higher Education to the Old Farm Trail," by Barbara George, synthesizes the restoration of a closed state mental hospital into an urban park of trails, meadows, and sports fields with the sustenance of a tenuous career in higher education. George, a doctoral candidate in composition and rhetoric at Kent State University, signed on as a facilitator at Boyce Mayview Park's Outdoor Classroom in Pittsburgh to augment her work as an adjunct instructor in English. As a facilitator with irregular hours, George led nature programs and hikes throughout the park, teaching local children about the geography, ecology, and history of southwestern Pennsylvania, where the scars of the steel industry—slag heaps, railways, and discarded machinery—remain visible and palpable. With references to the scholarship of Marilyn Cooper, Rob Nixon, Derek Owens, Rachel Tillman, and others, George wrestles with what it means to teach literacy—in particular, the literacy to question critically the institutions and systems in which we exist—in a working-class and industrial history landscape transformed into a pastoral space for a wealthy community, in the position of adjuncting as an outdoor educator in order to make ends meet adjuncting in higher education. Her work as an outdoor educator helps her to reflect on and revise her pedagogies as an academic in a way that makes the unsustainability and value of corporate higher education transparent and collective. The results of her changed consciousness are resonating in collaborations with other graduate students and adjuncts for university support, and in cross-disciplinary, cross-sector action for campus sustainability initiatives, even as the experiences of graduate students and the careers of adjuncts remain unsustainable on many levels.

The last two chapters of this collection discuss how recognizing, internalizing, and enacting Indigenous epistemologies in higher education highlight the severe limitations of Western ontologies and create spaces for a profound restoration of biodiversity, Indigenous languages, sustainable land stewardship, and vibrant human collaborations across lines of culture, age, gender, and nation-state. "Meeting across Ontologies: Grappling with an Ethics of Care in Our Human-More-than-Human Collaborative Work" is an inspiring reflection on the research of Bawaka Country, whose human participants are academics and Yolŋu community members Laklak Burarrwanga, Ritjilili Ganambarr, Merrkiyawuy Ganambarr-Stubbs, Banbapuy Ganambarr, Djawundil Maymuru, Kate Lloyd, Sarah Wright, Sandie Suchet-Pearson, and Paul Hodge. The collective has engaged in geography and environmental studies research in Bawaka Country, "an Indigenous Yolŋu homeland in Northeast Arnhem Land, Australia," since 2006, approaching their work from diverse perspectives as "an elder and as caretakers of Bawaka Country, as an Indigenous principal and as a senior teacher at a community school, as non-Indigenous academics at two universities, as Yolŋu running an Indigenous-led and owned cultural tourism business, as members of families, and as part of Bawaka Country." Underpinning their collaboration is the fundamental Yolŋu concept of *wetj*, a philosophy that joins all human and "more-than-human" beings within a system of interconnected kinship that is maintained "through obligations of attention, response and responsibility." This powerful, holistic, and life-sustaining Indigenous philosophy is chronically challenged, as the authors outline, by the compartmentalization, hierarchies, research protocols, and funding cycles of the neoliberal university system. As well, non-Indigenous members of Bawaka Country (a name the authors now use for themselves when they publish, to recognize and honor that these Yolŋu homelands are the author of their work as much as the humans who transcribe the language of the land) still feel young in their understandings of the sophisticated depth of the Yolŋu epistemologies that inform their work, and strive to overcome these "blockages" in their own thinking. First-person narratives by members of Bawaka Country intersperse the collective's description of the empowering joys and the frustrating barriers they and their families have faced in "co-becoming" and sustaining together. Through the "ethics of care" embedded in *wetj*, Bawaka Country tells a story of the kinship networks that may be formed by community-based research educators and Indigenous peoples, "with potentially profound implications for intergenerational self-determination, sustainability, and well-being."

Finally, Margaret Noodin's essay "*Ganawendamaw*: Anishinaabe Concepts of Sustainability" articulates how her teaching of Anishinaabe language and literature necessarily educates for sustainability, because Anishinaabe worldview as encoded in language—similar to the Yolŋu concept of *wetj*—expresses an epistemology in which all aspects of the natural world exist in a relationship of mutuality. While the sustainability of Anishinaabe language and culture itself is threatened on many fronts, Noodin nevertheless asserts that "teaching endangered concepts of sustainable ecological relationships is best done using endangered languages and literatures." As a poet, linguist, educator, and Anishinaabe community member, Noodin describes how the English word "sustainability," which "connotes a passive object that needs attention and is subject to an outside agency for its maintenance," contrasts sharply with "the Anishinaabe term '*ganawendamaw*' [which] is a verb that connotes a spectrum of animacy for all life allowing rocks, water, and humans to be described as co-equal partners in the creation, maintenance, and evolution of a place." Through teaching that all aspects of the natural world are relatives rather than resources in Anishinaabemowin (Anishinaabe language), Noodin's Anishinaabemowin and American Indian studies classes at the University of Wisconsin–Milwaukee allow students to witness the deep and dynamic place-based understandings arising from the people's origin stories and transmitted through generations of Anishinaabe storytellers and speakers of the language. Noodin's discussion of Anishinaabemowin stories highlights how they "demonstrate principles of adaptive learning and social-ecological resilience" that have long sustained Anishinaabe peoples. Despite ongoing challenges to the maintenance of Indigenous languages and cultural practices, Noodin's chapter reveals how her pedagogy of interwoven Anishinaabe language and stories at once sustains her people's culture and herself as a teacher and cultural worker.

▪ ▪ ▪

The questions raised by authors in this collection, and perhaps even some of the answers they find to these questions, may not be especially new. Yet we feel that asking the authors in these pages to include personal reflections about where, how, and why they teach for sustainability—in places they mostly aren't from, and in which they may or may not remain, whether or not by choice—gives voice to a community of educators grappling with similar issues in a range of far-flung places and adds an important dimension to existing texts in sustainability education. Despite the diversity of academic experiences in these pages, many of these authors' stories are uncannily similar, and we believe sustainability educators in a range of academic disciplines and geographies will also finds shades of their own higher-education stories here.

Of course, the challenges of sustainability in higher education are not just the stories made visible in these pages. Rather, they are the stories of those who are not included here as well—the stories that authors initially hoping to be involved with this collection were ultimately unable to contribute, for some of the very reasons discussed here about unsustainable environments in higher education. One group of faculty planned to write about their university's closure of its interdisciplinary environmental research center, but institutional politics and the dispersal of its faculty prevented them from finishing their chapter. Another university's revision of standards for faculty evaluation rejects narratives or reflections on pedagogy for the purposes of promotion and tenure, leading another would-be contributor to withdraw her planned chapter. Even as some universities support efforts to bring sustainability into their classrooms, some undermine such efforts by featuring climate-change deniers as commencement speakers while cultivating them as potential donors. At other institutions, faculty who excel in teaching, research, and service and are promoted to administration find themselves unable to bring about the change they know they can help shepherd, for the more their ideas for assessment and accountability take into consideration the real world of learning and the experiences of faculty and students, the more likely they are to be marginalized, dismissed, or forced out by their own administrations. All these are the stories of unsustainability in higher education in our time, and the academy is the poorer for the realities of these experiences.

Simultaneously, we find at the writing of this preface increasingly unsustainable political and social environments both within the United States and globally. It is jarring to read that 2015 was the deadliest year on record for "environment defenders," according to Global Witness, with about three killings a week in 2015 of individuals defending their lands, forests, or waterways. To the National Association of Scholars, sustainability is a fundamentalism that limits freedoms on the basis of fearmongering. In so many places, governors and legislatures are cutting funds for public higher education, while the U.S. Black Lives Matter movement has revealed the imbalance of power that is the result of longstanding, systemic racism. Worldwide, we see the ravages of instability that exacerbates unsustainability—just as headlines today proclaim that climate change is unfolding at a rate even worse than scientists predicted, more catastrophic than they thought. Mass migrations by populations fleeing violence are now met with xenophobia and violence in their countries of asylum, while U.S. President Donald J. Trump moves forward on building border walls and retreating into belligerent isolationism.[6]

Such destructive ideologies and fascistic practices are clearly antithetical to healthy and sustainable human and more-than-human generations. "If our collective future is to be harmonious and whole," Gregory Cajete writes, "*or* if we are even to have a viable future to pass on to our children's children, it is imperative that we actively envision and implement new ways of educating for ecological thinking and sustainability. The choice is ours, yet paradoxically we may have no choice."[7] As the chapters in this collection make clear, to educate for sustainability in the context of not only environmental devastation but violence and brutality toward human beings requires maintaining a critically reflective perspective, a calm head, an open heart, and ongoing education through connecting with one another as thoughtful professionals and as compassionate human beings.

NOTES

1. United Nations, *Our Common Future* (Oslo: World Commission on Environment and Development, 1987), I.3.27, http://www.un-documents.net/our-common-future.pdf; Jeannette C. Armstrong, "En'owkin: Decision-Making as if Sustainability Mattered," in *Ecological Literacy: Educating Our Children for a Sustainable World*, ed. Michael K. Stone and Zenobia Barlow (San Francisco: Sierra Club Books, 2005), 12.
2. Stuart Pearson, Steven Honeywood, and Mitch O'Toole, "Not Yet Learning for Sustainability: The Challenge of Environmental Education in a University," *International Research in Geographical and Environmental Education* 14, no. 3 (2005): 174, http://dx.doi.org/10.1080/10382040508668349; Chris Park and Michael Allaby, "Ecological Sustainability," in *A Dictionary of Environment and Conservation* (New York: Oxford University Press, 2013); Park and Allaby, "Environmental Sustainability," in *A Dictionary of Environment and Conservation*; Park and Allaby, "Sustainable Use," in *A Dictionary of Environment and Conservation*; Park and Allaby, "Sustainable Management," in *A Dictionary of Environment and Conservation*; Michael Allaby, "Sustainability," in *A Dictionary of Ecology* (New York: Oxford University Press, n.d.); Iain McLean and Alistair McMillan, "Sustainable Development," in *The Concise Oxford Dictionary of Politics* (New York: Oxford University Press, 2009); Andrew Brennan, "Development Ethics," in *The Oxford Companion to Philosophy* (New York: Oxford University Press, 2005); Kent E. Portney, *Sustainability* (Cambridge, MA: MIT Press, 2015), 194.
3. Jeremy L. Caradonna, *Sustainability: A History* (New York: Oxford University Press, 2014), 4, 5. In this comprehensive overview, Caradonna traces the roots of sustainability to the late seventeenth century, surveys its development and dissemination, and assesses the challenges that confront it.
4. Derek Owens, *Composition and Sustainability: Teaching for a Threatened Generation* (Urbana, IL: National Council of Teachers of English, 2001), xi; Fred Gale et al., "Four Impediments to Embedding Education for Sustainability in Higher Education," *Australian Journal of Environmental Education* 31, no. 2 (2015): 248–249, https://doi.org/10.1017/aee.2015.36; Margaret Robertson, *Sustainability Principles and Practice* (London: Routledge, 2014), 308.
5. Richard M. Carp, "Toward a Resilient Academy," in *Higher Education for Sustainability: Cases, Challenges, and Opportunities from Across the Curriculum*, ed. Lucas F. Johnston (New York: Routledge, 2013), 234; Pearson et al., "Not Yet Learning for Sustainability," 183.
6. Alice Harrison and Billy Kyte, "2015 Sees Unprecedented Killings of Environmental Activists," Global Witness, June 20, 2016, https://www.globalwitness.org/en/press-releases/2015-sees-unprecedented-killings-environmental-activists/; Rachelle Peterson

and Peter H. Wood, *Sustainability: Higher Education's New Fundamentalism*, National Association of Scholars, March 2015, https://www.nas.org/images/documents/NAS-Sustainability-Digital.pdf.

7. Gregory Cajete, *Look to the Mountain: An Ecology of Indigenous Education* (Skyland, NC: Kivaki Press Inc., 1994), 23.

BIBLIOGRAPHY

Applewhite, James. *River Writing: An Eno Journal*. Princeton, NJ: Princeton University Press, 1988.

Armstrong, Jeannette C. "En'owkin: Decision-Making as if Sustainability Mattered." In *Ecological Literacy: Educating Our Children for a Sustainable World*, edited by Michael K. Stone and Zenobia Barlow, 11–17. San Francisco: Sierra Club Books, 2005.

Cajete, Gregory. *Look to the Mountain: An Ecology of Indigenous Education*. Skyland, NC: Kivaki Press Inc., 1994.

Caradonna, Jeremy L. *Sustainability: A History*. New York: Oxford University Press, 2014.

Carp, Richard M. "Toward a Resilient Academy." In *Higher Education for Sustainability: Cases, Challenges, and Opportunities from Across the Curriculum*, edited by Lucas F. Johnston, 223–237. New York: Routledge, 2013.

Gale, Fred, Aidan Davison, Graham Wood, Stewart Williams, and Nick Towle. "Four Impediments to Embedding Education for Sustainability in Higher Education." *Australian Journal of Environmental Education* 31, no. 2 (2015): 248–263. https://doi.org/10.1017/aee.2015.36.

Owens, Derek. *Composition and Sustainability: Teaching for a Threatened Generation*. Urbana, IL: National Council of Teachers of English, 2001.

Pearson, Stuart, Steven Honeywood, and Mitch O'Toole. "Not Yet Learning for Sustainability: The Challenge of Environmental Education in a University." *International Research in Geographical and Environmental Education* 14, no. 3 (2005): 173–186. Http://dx.doi.org/10.1080/10382040508668349.

Portney, Kent E. *Sustainability*. Cambridge, MA: MIT Press, 2015.

Robertson, Margaret. *Sustainability Principles and Practice*. London: Routledge, 2014.

United Nations. *Our Common Future*. Oslo: World Commission on Environment and Development, 1987. Http://www.un-documents.net/our-common-future.pdf.

PART 1.

Confronting the Challenges of the Places We Are

Unrooted: Dislocation and the Teaching of Place

Jennifer L. Case

> Whether and in what sense I experience a particular location as a "place" will be further affected by such factors as how rooted or peripatetic my previous life had been, what kinds of surrounding I am conditioned to feel as familiar or strange, and so forth. So place-sense is a kind of palimpsest of serial place-experience.
>
> —Lawrence Buell, *The Future of Environmental Criticism*

My first spring as a graduate student in English, thirteen other graduate students and I took a seminar on environmental criticism. We met in a basement classroom at the University of Nebraska-Lincoln, a land grant university located next to the banks, businesses, and restaurants of the capital city's downtown. Though UNL is known best for its football team—and the football stadium becomes the state's third largest city on game day—our classroom had little sense of pretense. The walls were off-white, the fluorescent lights flickered, and the desks were too small for our books. From the tiny window high on the back wall, a faint light struggled to make its way past the shrubs and the footsteps of pedestrians. As my colleagues and I congregated in the still-dark evenings of early spring, what we returned to, more than anything else during pre- and post-class discussions, was the subjectivity of our environmental perceptions and, more specifically, our perceptions of beauty in nature. The draft from the lone window made us pull on sweaters and jackets, an act that invariably led us to complain about the weather. We'd just read Lawrence Buell's *The Future of Environmental Criticism: Environmental Crisis and Literary Imagination* (2005), and we all felt nostalgic. The students from Alabama and Georgia said it was too cold in Nebraska. When the wind blew, as it did then, they couldn't breathe. In response, those of us from farther north belittled Nebraska's winters. The flakes that fell and melted within two days, we told the group, weren't snow. They were dandruff. Real snow would stay all winter, forming mini-mountains in the corners of parking lots. And as for the landscape: depressing. The woman from upstate New York missed trees; I said that I missed the lakes and the rivers that surrounded all the cities I'd lived in before moving to Nebraska. Only the sole Nebraska native defended treeless environments. Trees, she said, made her claustrophobic. She said she'd like to take a chainsaw to them all.

I mention these conversations not only because they demonstrate how much place affects us (though this, arguably, was a group predisposed to an interest in sustainability and place-based scholarship), but also because they demonstrate how "place" plays out in the academic experience. Though the demographics of the graduate students and faculty at any particular university will vary, one could plausibly say that very rarely will everyone come from the same place. Academia encourages us to move around—to new cities, new counties, new states—sometimes even new countries. This reality means that at universities such as UNL, each year graduate students and faculty come together from multiple regions of the United States and from multiple countries in the world. The university is new. The place is new. The bioregion, the flora, and the climate are new. We, effectively, have been uprooted, and as we acclimate to a particular program, we must also acclimate to the culture and nature of that particular city and state.

Although I am most interested in the ways this dislocation affects the ability of composition instructors—specifically graduate students and contingent faculty—to engage in sustainability education and teach about place, the paradox has in fact served as the impetus of much place-based thinking. Many place-based compositionists and educators begin their treatises by acknowledging the placelessness of academia. Robert Brooke introduces *Rural Voices: Place-Conscious Education and the Teaching of Writing* (2003) by describing the displacement he felt as an academic, moving from his homeplace in Denver, to the eastern United States for school, and then to eastern Nebraska to teach. Laird Christensen and Hal

Crimmel open their edited anthology *Teaching about Place: Learning from the Land* (2008) by recognizing their own geographical nomadism:

> Neither of us inhabits our native places: The one raised north of New York's Adirondacks now lives along Utah's Wasatch Front. The one from the Columbia River's southern shore now lives in Vermont, near the head of Lake Champlain.
>
> This collection is born of that displacement, a response to our decades spent chasing education, adventure, or employment. The tension between the appeal of the road and our desire for roots has defined our lives, personally and professionally, and so we find that our complicated relationship with place informs our teaching and our writing.

Outside of place-based composition, scholars of sustainability education more broadly conceived have made similar statements. In *Ecological Literacy: Educating Our Children for a Sustainable World* (2005), environmental educator David Orr writes that the typical college disciplines don't address "the art of living well in a place," a fact that influences the kinds of lives academics lead: "Place is nebulous to educators because to a great extent we are a deplaced people for whom our immediate places are no longer sources of food, water, livelihood, energy, materials, friends, recreation, or sacred inspiration."[1] Such comments, in fact, are common in introductions and personal anecdotes across the interdisciplinary field, pointing to a general dissatisfaction with the lifestyle and geographical placelessness contemporary academia seems to call for—a placelessness that affects both students and local communities.

Indeed, it is this disconnect between academia and local communities that propels much scholarship on place-based pedagogy, whether in environmental education or first-year composition. The educational system, many place-studies scholars argue, now emphasizes state and national standards to a degree that overshadows local knowledge and threatens local cultures. "The local places that students and staff and faculty go home to after leaving the university behind," Derek Owens contends, "remain largely invisible, supposedly unrelated to the activity of the academy, despite mission statement rhetoric about serving community and helping students become responsible citizens." Similarly, Eric Ball and Alicia Lai argue in "Place-Based Pedagogy for the Arts and Humanities" (2006) that the educational system privileges a "(trans)national agenda" that ignores and does actual harm to local areas. It encourages displacement by "failing to cultivate care for places among students" and, instead, "providing students with a credential that enables pursuit of careers that do actual damage to local communities"—thus weakening the communities from which students come and in which instructors teach.[2] In other words, academia's emphasis on the cognitive has worked to harm local regions—not only by overlooking their significance but also by physically luring students away.

The latter concern remains particularly important in highly rural or undervalued regions, such as Nebraska. Place-based scholars and writers such as John Price see place-based writing as a means to "counter the culturally informed notions of the grasslands as a place to move through and beyond." Price does not want to see his students move from Nebraska in search of economic opportunities in Chicago. He does not want academia to endorse a worldview that would deter cultural, artistic, and economic sustainability in the Great Plains states. Instead, he works to help students appreciate the places they come from and recognize the ways they could use their education to improve, or at least participate actively in, those communities. Moreover, many place-based compositionists argue that focusing on the local, especially in rural or undervalued areas, can address institutional learning goals better than the place-blind practices that university systems and national educational agencies have typically promoted. Brooke's work for the Nebraska Writing Project and the Rural Institute has done much to establish place-based composition as a noteworthy pedagogy in rural areas, and the following passage from his *Rural Voices* offers an especially rich explanation of place-based composition's key goals and tone:

> Learning and writing and citizenship are richer when they are tied to and flow from local culture. Local communities, regions, and histories are the places where we shape our individual lives, and their economic and political and aesthetic issues are every bit as complex as the same issues on national and international scale. Save for the few of us who become senators and CEOs and *National Geographic* reporters, it is at the local level where we are most able to act, and at the local level where we are most able to affect and improve community. If education in general, and writing education in particular, is to become more relevant, to become a real force for improving the societies in which we live, then it must become more closely linked to the local, to the spheres of action and influence which most of us experience.[3]

This sense of local communities, whether urban or rural, as a rich site for inquiry—one that can help prepare students to be better citizens—is a prominent thread across the place-based pedagogies that have emerged in writing classrooms at the university level.

Interestingly enough, most of the graduate students in my ecocriticism seminar in the basement classroom at UNL were not particularly linked to our own local community but instead exemplified Price's concerns about a lack of place attachment in the "heartland." Though we shopped at farmers' markets, participated in community gardens, and trekked to the Platte River to observe the annual Sandhill Crane migration, we did not change our cell phone numbers or consider Nebraska "home." We did not know the names of the elected leaders who represented our neighborhoods. We had come to Nebraska primarily for a graduate degree; we did not expect to stay long. Even more telling: although we recognized that any geographical preferences would severely limit us during the job search, few us of would choose to remain in Nebraska. Our dream jobs were elsewhere.

Indeed, it continued to surprise me that I was trying to teach place-based courses that emphasized sustainability when I felt little personal attachment to Lincoln or the landscape surrounding it. The endless fields of corn I drove past when moving to Lincoln and the flat, concrete landscape of the city (where what I would call a "ravine" was labeled a river) appealed to me little. I sighed in relief whenever I left the state for holidays, and I slouched in my car as I drove back toward the brown plains. In essence, I manifested the central concern of most scholars in the discourse: disinterest in the material environment. To paraphrase Nedra Reynolds, I had been conditioned to take space for granted and instead focus on the capitalistic gains of my graduate education. It became easy for me, then, to separate myself from the environment. Although I recycled everything I could, shopped at the local co-op, and walked or biked to the university rather than drive, I did this blindly and out of habit—not out of a concern for the Nebraskan landscape or the ecological problems the region faced. The place was negligible; I ignored how the classrooms I taught in, the city streets I commuted on, and the farmland and prairie I hiked near affected and were affected by discourse. In the words of Price, I, like so many instructors, struggled "to find a reason to care."[4]

Nonetheless, my commitment to the place-based pedagogies promoted by Brooke, Jonathan Mauk, Orr, Owens, and Price, and my belief that place-based writing can help students find their footing in academic discourse could not help but become part of my classrooms. As a result, I began my freshmen composition courses the following year with a personal narrative on place. The assignment asked my students to write about a place that was particularly memorable or significant to them. By asking my students to consider how that place continued to affect them—how their identity or personality or history had been shaped by the materiality of the place and their experiences of the place—I attempted to defamiliarize them from their surroundings, to help them recognize, as Owens has said, "that who we are and what we have to say is in so many ways interwoven, directly and indirectly, consciously and unconsciously, with our local environs." Certainly, my reasoning then was (and today remains) ideologically based. I designed the course around "place" because of my interest in place-based composition and my belief that helping students locate themselves within a place would not only give them more ownership/authority of their work, but also make it easier for them to become active members in the writing community. I shared Owens's concern for sustainability; I wanted my students to care about the health of their communities. Aside from that desire, however, I also, like Mauk, recognized the value of place-based writing as a means of validating student experience and integrating academic thinking with that experience.[5] As first-year students transition into college, place-based writing allows them to carry with them the knowledge and authority they have acquired, thus helping them situate themselves within the academy and develop their own senses of agency.

This emphasis on agency, I found, gave my students courage as writers. My students responded positively to the place-based personal narrative; they were excited to write about hometowns, family cabins, favorite vacation spots, or relatives' houses. Even more important: they thought about their places in sophisticated ways and recognized how those places shaped them. For instance, Javaun used the project to explore his childhood in inner city St. Louis. Though he recognized that he came from a rough, sometimes violent environment, where many of his classmates would not want to live, he also embraced the lessons he had learned from his cousins and neighbors. His family, he said, enjoyed each other's company even though they did not have many material comforts, and he hoped to replicate that value in his own life. Another student, Michael, wrote about the golf course he worked at while attending high school. Because Michael's family moved each year from military base to military base, high school was the first time he stayed put for more than two years. In writing about the golf course, Michael realized that it was this emplacement, just as much as the coworkers he became friends with, that made the golf course so important to him. Such projects and realizations quickly validated my reasons for designing the course as I had. Students, by investigating the places they had come from and the related societal and community issues present in those places, gained the confidence and authority they needed to enter into the more complex discussions that surrounded them.[6]

As the semester progressed, we tackled the other genres common to first-year composition courses: the rhetorical analysis, the argument, and the research essay. With each, I again asked the students to ground their topics in issues important to the communities and places they knew well. Although I didn't require students to write about the same place the entire semester, many did. A student who used the personal essay to write about her high school later found it empowering to analyze and create arguments related to her school's referendum. A political science major who wrote about his grandparents' farm later researched how political leanings differed in rural and urban areas. An ecology major whose "significant place" was the Henry Doorly Zoo and Aquarium in Omaha contrasted the zoo's conservation policies with policies promoted by other environmental agencies.

As a result, I didn't wade through stereotypical essays on abortions, the death penalty, or stem-cell research—polarized topics entirely devoid of the students' voices, interests, and senses of agency (though those topics, too, can be approached effectively through a place studies perspective). Rather, I read essays that clearly linked with the students' current interests and future career goals. In this way, we managed to build a platform from which they could connect their individual lives with the livelihoods of their local communities, as well as with national and global concerns.

I have since taught numerous place-based composition courses, at UNL, Binghamton University in upstate New York, and, most recently, the University of Central Arkansas. At each location—whether I worked with students from rural Nebraska, the urban Midwest, the South, or the boroughs of New York—my students have embraced a place-based approach. My students in New York tended to write about more cosmopolitan and global subjects: the subway station near an apartment complex in New York City, the airport in the Dominican Republic where they greeted their extended family during summer visits home, a dumpling shop near a grandmother's house in Shanghai, China, and a street in Queens that was quickly changing because of gentrification. They also were less resistant to the idea that those locations had value in the larger world. New York City and its surrounding areas hold more weight in the dominant cultural narrative than a small Nebraskan farming town, after all. Yet the enthusiasm I've encountered at each institution is the same. Indeed, many of my students hope to return to their hometowns after college. As a result, they enjoy writing about their high schools and old neighborhoods, and they often conclude their essays with the hope that, eventually, they will settle down somewhere similar. When I read these essays, I often recognize myself in them. My students write about their significant places with the same nostalgia that my fellow graduate students and I once voiced in that dreary basement classroom during our ecocriticism course—a nostalgia that sometimes interferes with my students' (and my colleagues') capacities to commit to new places. However, unlike my students, I do not see myself returning to my own hometown anytime soon. The characteristics of academia make homecoming rather difficult.

When I first went on the job market, I received the same warning and advice that hundreds of colleagues before me had: if you have a geographical preference, your chances of landing a tenure-track job get smaller and smaller as that geographical area shrinks. PhD graduates entering the job market are encouraged to apply to nearly anything and everything that they can—to conduct a national search. Even more notable, however, is that even national searches may not lead to, and increasingly aren't leading to, tenure-track positions. According to the 2014 *Report on the MLA Task Force on Doctoral Study in Modern Language and Literature*, doctoral recipients before 2008 could expect about as many tenure-track openings a year as doctoral degrees awarded: one thousand. After 2008 that shifted, and today there are only about six hundred tenure-track openings a year, but still one thousand new graduates. This chasm has caused MLA itself to use the word "bleak" when describing the job market.[7] To put it bluntly, if I wanted to end up with a stable job in a "place" that resonated with me, I shouldn't get my hopes up. Rather, I needed to apply broadly and be prepared to accept what came. Even if it wasn't in an area of the country that appealed to me. Maybe even if it wasn't a long-term, full-time position.

As most professionals in academia are well aware, the percentage of courses run by graduate students and contingent faculty has grown exponentially. For a variety of reasons, economic and otherwise, colleges and universities—from two-year to four-year institutions—are relying more and more on graduate students and adjuncts to teach undergraduate courses, especially the core courses. The problem is receiving attention and discussion, with national dialogue sparked by the 2015 National Adjunct Walkout and calls for an adjunct minimum wage, but the fact remains: today, 76.4 percent of college and university instructors are adjuncts and contingent faculty. At public doctoral institutions, graduate assistants make up 31.7 percent of the academic work force. Another MLA publication, "Where Are

They Now? Occupations of 1996–2011 Ph.D. Recipients in 2013," reveals that when those grad assistants graduate, only about half of them will end up on the tenure track. The rest will leave academia to pursue work in the for-profits and nonprofits, or teach in a limited-term capacity.[8] For many, this reality can mean teaching at more than one institution at once, or hopping from visiting assistant professor contract to visiting assistant professor contract. In other words, a large percentage of the academic work force that could be—and we'd hope would be—teaching sustainability or place-based courses struggles with displacement within the college and university setting.

The complications of this displacement become even more clear—and ironic, when it comes to the teaching of place—in the context of job satisfaction. In "Supporting the Academic Majority: Policies and Practices Related to Part-Time Faculty's Job Satisfaction," published in the *Journal of Higher Education* (2015), M. Kevin Eagan Jr., Audrey J. Jaeger, and Ashley Grantham outline the challenges that contingent faculty face: "limited or no office space, a lack of clerical or administrative support, restricted involvement in campus governance, and no guarantee of continued appointment." Such faculty, they summarize, tend to feel isolated, like second-class citizens. The feeling of isolation is even greater among contingent faculty whom Eagan, Jaeger, and Grantham identify as underemployed—those adjunct faculty who would prefer full-time employment, rather than those adjunct faculty who have chosen to teach part-time for other reasons. Using a 2010–2011 HERI Faculty Survey, Eagan, Jaeger, and Grantham identify that 73 percent of part-time faculty, in fact, consider themselves underemployed. This means that, considering contingent faculty comprise 75 percent of the academic work force, nearly 55 percent of the nation's college and university instructors would prefer a more stable, full-time position. More than half of the nation's instructors do not feel grounded in their university setting. This displacement can also have significant impacts on job performance and satisfaction. Underemployment has been linked to "decreasing organizational commitment and citizenship"; it has also been shown to negatively affect mental and physical health.[9] In other words, underemployment can lead to exactly the kinds of thinking—on an even more personal and profound level—that place-based philosophies, from the start, have sought to counter: disinterest in the material environment and a lack of care for the local community.

As an instructor of first-year composition who has taught as both a graduate assistant and an adjunct, then, I wonder how much this displacement affects our abilities as place-based educators. Place-based scholarship has carefully explored the benefits of place-based pedagogy for students. But what unique challenges and opportunities does place-based writing present for composition instructors, especially graduate teaching assistants and limited-term faculty, who will likely reside in a region for little more than half a decade? For a limited-term faculty member who might not have a university computer and who likely does not have an office of his or her own—if an office at all? For the limited-term faculty member who carries textbooks back and forth between a temporary apartment and the campus, who can count on a contract for only a single semester? For the faculty member who moved across the country for a one-year visiting assistant professor contract and thus is constantly eyeing job advertisements and sending out applications? For the contingent faculty member who has taught at a single university for more than twenty years but still has no say in the department's governance? In other words, if first-year writing offers a unique opportunity to integrate sustainability into education—to make the significance of place and students' impacts on those places a part of their educational experience—what does it mean if the instructors who tend to teach these courses are also people in need of rooting, of placement? We cannot expect our faculty members to be able to teach sustainability if they themselves are balanced precariously in unsustainable environments. And if we do want these faculty members to educate for sustainability—because we desire a university system that will prepare students to tackle, with creativity and innovation, the environmental and societal issues their local, national, and global communities will face—the question then becomes: How can we help these faculty members and motivate them? How can we support a faculty member with as many—or perhaps even fewer—emotional ties to the university as the incoming freshmen he or she is teaching? What would propel a new graduate student or contingent faculty member to *care* about the local environment he or she is likely quite new to, especially if it is an academic institution in which he or she has very little stake?

I do not yet have answers to all of these questions. Maybe I never will. The national trend towards relying on contingent faculty members, for instance, is far too complex—and outside of my expertise—to solve here in a single chapter. Yet they are questions that we need to be asking, especially as we seek to engage and problematize sustainability within higher education. Thus, any discussion of educating for sustainability must take into account the fact that academia itself, in its current form, produces an unsustainable environment. The inherent mobility of academic life often does not, at least initially, facilitate a supportive, sustainable relationship with place, and increasingly, the jobs and working conditions have been proving unsustainable, too. These actualities have significant repercussions when it comes to sustainability education. Simply put, the professors and instructors who can and should be educating for sustainability cannot if they are working in unstable positions. Instead, they are worried about their financial futures and struggling to adjust to new and changing environments. We must recognize this emotional toll if we want to support sustainability education. We must recognize that graduate students and contingent faculty members are just as much in need of "placement" as the undergraduate students they teach—the first-year students arriving with futons strapped to pickup trucks, as well as the commuter students struggling to balance school with a full-time job. They all need to learn a new terrain. They all, faculty and students alike, would benefit and gain satisfaction from a greater sense of authority and ownership within the institutional, academic, and often even the geographic space.

As a young academic, I have become used to dislocation. I moved from Minnesota to Nebraska, and from Nebraska to upstate New York. After earning my PhD, I adjuncted for a year before moving once again, this time to Arkansas for my first full-time position. I have learned to recognize a pattern: it takes at least a year for me to feel grounded. For a year (and maybe longer), I do not feel connected to the program, the city, or the state. Though the buildings I teach in have histories, I am not part of those histories, and when my first-year students from Nebraska, New York, and Arkansas discuss what it is like to grow up in their regions, I stand in front of them with only stories of Minnesota snow. In those moments, my own, migratory background identifies me as a globetrotter, a wanderer, and not exactly a good example for the kind of placeness I'd like to see my students embrace in their own lives. Indeed, as I try to validate my students' backgrounds and integrate place-based assignments and readings, I often struggle to have authority or voice when I know so little about the complex history of hydrofracking in New York, or Southern evangelical culture in Arkansas. At each and every institution, I need to learn, and that learning takes time. Teaching sustainability, I have accepted, requires purpose and care, and staying motivated to integrate place-based practices—especially if I know I will likely move again soon—benefits greatly from joy.

As a result, my greatest endorsement of place-based writing assignments is this: I have always enjoyed reading them. In Nebraska, as a new graduate student in a new state, I liked learning about the seventy-person towns my students grew up in, the hunting trips they went on with their families, the strip malls in Omaha they visited each weekend during high school, and the coffee shops they frequented in order to study and laugh with friends. As we discussed place—as I encouraged them to think in new, more sophisticated ways about the places they were familiar with and the temporary positions we shared in the university—we began a mutual inquiry into the region. I helped defamiliarize that area for them, while they, in turn, helped me to care about the cornfields and the landscape I earlier considered ugly. As they shared their stories and began to explore *how* culture is intertwined with the rural and urban landscapes they'd often overlooked, I began to better appreciate the texture and nuances of their backgrounds. Certainly, the practice had pedagogical foundations. By starting with something the students could claim ownership of—personal experiences of places significant to them—I attempted, as Eric Ball and Alice Lai describe, to allow them "a greater voice in the spatial politics of culture."[10] I attempted to validate their experiences and concerns even as we entered into more critical discussions of place, culture, and audience. However, the practice also had real benefits for me. The flat farmland I had once sped past on the interstate was now the fourth-generation homestead of a student debating the pros and cons of ethanol. An Indian reservation my colleagues and I drove past en route to a conference transformed into the hometown of a student concerned with liquor sales just outside the reservation's borders. And that coffee shop I'd visit on weekends for cherry crumb cake? It's where a local student performed during open mic nights and discovered her desire to support arts education in the state's schools. In Nebraska—and later New York and Arkansas—I, like my students, learned to "place" myself within the environment—within that particular university, that particular city, and that particular bioregion. In the intertwining of our stories, I began to develop roots.

NOTES

1. Robert E. Brooke, ed., *Rural Voices: Place-Conscious Education and the Teaching of Writing* (New York: Teachers College Press, 2003), 1–5; Laird Christensen and Hal Crimmel, "Introduction: Why Teach about Place?" in *Teaching about Place: Learning from the Land*, ed. Christensen and Crimmel (Reno: University of Nevada Press, 2008), ix; David Orr, *Earth in Mind: On Education, Environment, and the Human Prospect* (Washington, DC: First Island Press, 1994), 126.
2. Derek Owens, *Composition and Sustainability: Teaching for a Threatened Generation* (Urbana, IL: National Council of Teachers of English, 2001), 70; Eric L. Ball and Alice Lai,

"Place-Based Pedagogy for the Arts and Humanities," *Pedagogy* 6, no. 2 (Spring 2006): 263.

3. John Price, "Idiot out Wandering Around: A Few Words about Teaching Place in the Heartland," in Christensen and Crimmel, *Teaching about Place*, 149; Brooke, *Rural Voices*, 4–5.
4. Nedra Reynolds, "Composition's Imagined Geographies: The Politics of Space in the Frontier, City, and Cyberspace," in *Relations, Locations, Positions: Composition Theory for Writing Teachers*, ed. Peter Vandenberg, Sue Hamm, and Jennifer Clary-Lemon (Urbana, IL: National Council of Teachers of English, 2006), 234; Price, "Idiot out Wandering," 141.
5. Owens, *Composition and Sustainability*, 37; Jonathan Mauk, "Location, Location, Location: The 'Real' (E)states of Being, Writing, and Thinking in Composition," in Vandenberg, Hamm, and Clary-Lemon, *Relations, Locations, Positions*, 222.
6. In this chapter, students are identified pseudonymously.
7. *Report of the MLA Task Force on Doctoral Study in Modern Language and Literature* (New York: Modern Language Association of America, 2014), 5, https://www.mla.org/content/download/25437/1164354/taskforcedocstudy2014.pdf.
8. John W. Curtis, *The Employment Status of Instructional Staff Members in Higher Education, Fall 2014* (Washington, DC: American Association of University Professors, 2014), 8, https://www.aaup.org/sites/default/files/files/AAUP-InstrStaff2011-April2014.pdf; *Report of the MLA Task Force*, 20; "Where Are They Now? Occupations of 1996–2011 PhD Recipients in 2013," *The Trend: The Blog of the MLA Office of Research*, February 17, 2015, https://mlaresearch.commons.mla.org/2015/02/17/where-are-they-now-occupations-of-1996-2011-phd-recipients-in-2013-2/.
9. M. Kevin Eagan Jr., Audrey J. Jaeger, and Ashley Grantham, "Supporting the Academic Majority: Policies and Practices Related to Part-Time Faculty's Job Satisfaction," *Journal of Higher Education* 86, no. 3 (May–June 2015): 449, 463, 455.
10. Ball and Lai, "Place-Based Pedagogy," 280.

BIBLIOGRAPHY

Ball, Eric L., and Alice Lai. "Place-Based Pedagogy for the Arts and Humanities." *Pedagogy* 6, no. 2 (Spring 2006): 261–287.

Brooke, Robert E., ed. *Rural Voices: Place-Conscious Education and the Teaching of Writing*. New York: Teachers College Press, 2003.

Buell, Lawrence. *The Future of Environmental Criticism: Environmental Crisis and Literary Imagination*. Malden, MA: Blackwell, 2005.

Christensen, Laird, and Hal Crimmel, eds. *Teaching about Place: Learning from the Land.* Reno: University of Nevada Press, 2008.

Eagan, M. Kevin, Jr., Audrey J. Jaeger, and Ashley Grantham. "Supporting the Academic Majority: Policies and Practices Related to Part-Time Faculty's Job Satisfaction." *Journal of Higher Education* 86, no. 3 (May–June 2015): 448–483.

Mauk, Jonathan. "Location, Location, Location: The 'Real' (E)states of Being, Writing, and Thinking in Composition." In *Relations, Locations, Positions: Composition Theory for Writing Teachers*, edited by Peter Vandenberg, Sue Hamm, and Jennifer Clary-Lemon, 198–223. Urbana, IL: National Council of Teachers of English, 2006.

Orr, David. *Earth in Mind: On Education, Environment, and the Human Prospect.* Washington, DC: First Island Press, 1994.

Owens, Derek. *Composition and Sustainability: Teaching for a Threatened Generation.* Urbana, IL: National Council of Teachers of English, 2001.

Price, John. "Idiot out Wandering Around: A Few Words about Teaching Place in the Heartland." In Christensen and Crimmel, *Teaching about Place*, 139–153.

Reynolds, Nedra. "Composition's Imagined Geographies: The Politics of Space in the Frontier, City, and Cyberspace." In Vandenberg, Hamm, and Clary-Lemon, *Relations, Locations, Positions*, 226–257.

By the Lumbee River with Chad Locklear's "Swamp Posse"

Jane Haladay and Scott Hicks

For most of its length, the Lumbee flows through swamps and woodlands, a shadowy world of half-seen creatures and movements, a world in which man has intruded, where nature can never be forgotten.

—Adolph L. Dial and David Eliades, *The Only Land I Know: A History of the Lumbee Indian*

"What's a 'toten'?"

We've just begun discussing the first piece of literature we teach in our Environmental Literature course, Chad Locklear's "Swamp Posse" (2006). It's inevitable that more than one student in the class will have this question about one of the Lumbee terms Locklear includes in his short story, and it's inevitable that more than one student in the class will know very well the meaning of this word. We pause for a beat, noticing the Lumbee students looking sidelong at each other with smiles playing on their faces.

"Well?" one of us finally says, tossing the question back to the class. "Does anyone know what a 'toten' is?"

Several voices speak at once. "It's like a spirit," says one student. "Or kinda like a presence," says another. "Like you know someone or something's there, or you see something, like a sort of sign," says a third.

Our students' responses confirm Lumbee bibliographer Glenn Ellen Starr Stilling's description that "totens" can be "a smell, sound, or vision indicating the presence of a spirit." This term appears in "Swamp Posse" when Vince, the story's narrator, reunites with his boyhood friends by the Lumbee River and explains to them why he's returned to Robeson County: to record oral histories of Lumbee elders for his master's thesis project. "I've been focusing a lot on the supernatural stories lately," Vince tells his old crew. Shifting to Lumbee English after realizing he is "talking like a 'white boy,'" he adds, "You know what I'm talking bout, like them ghost stories, and stories about spirit guides and totens, animals and conjurers."[1]

While this distinctive Lumbee term might seem peripheral to the main ecocritical ideas of the story (such as its setting on the Lumbee River, a national and state wild and scenic river;[2] the characters' relationship to the landscape; and the centrality of fishing, hunting, and farming in the characters' lives), totens provide an invaluable point of entry because they reconfigure students' understandings of environment and ecology as more than solely scientific concepts and relationships. Totens introduce students to an inspirited landscape in a particular local culture and epistemology, demanding alternative ways of understanding a human and more-than-human world in a highly localized context, one in which a human presence always already exists both inside and beyond the landscape. As the class's opening reading, "Swamp Posse" emphasizes the relationship between the physical environs of Robeson County—its farm fields, its swamps, its wide open highways and back roads, the Lumbee River—and its communal and familial social interconnections.

Put simply, privileging "Swamp Posse" as our first course text, with its intense, contemporary focus on a particularly regional setting, is our way of beginning to indigenize our students' thinking about the place in which they now exist, the University of North Carolina, Pembroke, in conversation with the many places from which they come. As we draw students into the language of Locklear's characters, we are drawing out the voices of the Lumbee students to whom the names, setting, languages, and events of "Swamp Posse" will be felt and known. At the same time, we are connecting students unfamiliar with Robeson County and the Lumbee people to the communities that are home to our university and the coastal-plain blackwater river and floodplain in which our university exists, an ecosystem of bald cypress and tupelos, turtles and water snakes, pileated woodpeckers and yellow prothonotary warblers. Hunkering down with this short story helps us frame the ecocritical questions of our entire literature course, questions that will loop back and resonate in myriad ways with every student in the class as we next engage a more global and theoretical text, Greg Garrard's *Ecocriticism* (2012).[3]

Yet at the same time that "Swamp Posse" vividly represents a sense of long-term habitation in a specific environment, it also starkly reveals Robeson County as a complex place where historical violence and trauma to both human and other living beings has diminished it over time as a sustainable environment on many levels. For "sustainability involves more than 'the environment'; it is equally interested in social sustainability (often summarized as well-being, equality, democracy, and justice) and sensible economics but, above all, in the interconnectedness of these domains."[4] Locklear's story of Vince and his posse, through the course of one dramatic night, pulses with the tensions of how the political, economic, social, and ecological inheritance of this part of southeastern North Carolina has undergone large-scale ecological, social, and economic depletion for generations, requiring ongoing cycles of recovery that often seem like one step forward and two steps back. As educators in this place, we strive to show students how community sustainability requires environmental, social, and economic consideration; to guide students to recognize the opportunities and confront the problems with the places they inhabit; to understand how these places both shape and are shaped by them; and to commit to understanding and action rather than nostalgia and inaction, as we all continue to witness radical environmental changes on a global level. Simultaneously, we have wrestled with—and continue to wrestle with—this same ambivalence, as we live and work in, but are not *from*, southeastern North Carolina. Like snagged fishing line in the black waters of the Lumbee River, Locklear's story artfully tugs at these tangled concerns about the sustainability of the places we're from, the place we are, and the places to which we can never fully return.

Putting down Roots, Confronting Historical Trauma

To a certain extent, "Swamp Posse" has proved to be productive not only in bringing our students in Environmental Literature to the place they have chosen for college; it has also helped us, their professors, root ourselves through story to the place we have now taught since 2006. We did not always open our Environmental Literature course with Locklear's story. Our course, like many a professor's course, has developed through trial and error, with a range of assignments that, in our case, has included a local foods potluck the first year we taught the course (replete with Kentucky

Fried Chicken, black-eyed peas, kudzu salad, and Kool-Aid) to a smorgasbord of environmental prose, poetry, and drama set only in North Carolina or written only by North Carolina authors. Our ultimate arrival at opening with "Swamp Posse" was, in a way, a sort of pedagogical and curricular homecoming by two educators whose homes had been elsewhere.

Jane's first impressions of the university in 2006 had very much to do with her feelings about the place *as* a place: It was different from any physical environment she was used to; it was ecologically unfamiliar. But more than these differences (at least the pine trees, although a different species, were familiar from her homeplace) were the cultural contrasts she felt: those she heard in unfamiliar regional speech patterns and dialects; those she perceived in the very different, conservative religiosity of rural southeastern North Carolina; those she saw in the flat and swampy terrain of the Sandhills region. She also noticed an underdeveloped transportation system that had her search-committee members driving her to and from campus down long, winding rural roads with no streetlights or signposts, explaining where they were heading with impossible-to-remember directions like "When you get to the brown church on the left, take the east fork," to a university that had no bus, no metro, no shuttles, no public transportation system in place at all. The community bookstore, the twenty-four-hour Kinko's? No dice. Coming from the University of California at Davis, Jane initially wondered what kind of college town she'd landed in. Where are the Jamba Juice, the sushi bars, the Starbucks? When she interviewed at UNCP in 2006, she came to learn that even the nearest Starbucks was a forty-five-minute drive away and that there was no Trader Joe's in the entire *state.* This was *not* civilization as she knew it.

Scott, by contrast, had heard quite a bit about life in this part of the state. As a North Carolinian who grew up in the Research Triangle before it topped so many *Forbes*, *Sperling's Best Places*, *Travel + Leisure*, and *Money* lists for quality of life, education, and innovation, before it became a darling of Richard Florida's concept of the creative class, Scott knew that things were different Down East, especially along the South Carolina border. He had heard people mock this region; worse, everyone knew it was where Michael Jordan's father was murdered in 1993, his body dumped in a swamp, identified only by his son's championship ring. As Robeson County residents told *GQ* in a story about James Jordan's murder, every North Carolinian knew not to stop to take a rest in Lumberton—it was "Hell's backyard," one woman told the magazine. Even earlier, in 1988, in a desperate attempt to get the state to take seriously claims of pervasive corruption inside Robeson County's sheriff's department and court system, Native activists Eddie Hatcher and Timothy Jacobs, shotguns in hand, took over the local newspaper in 1988 and held nearly two dozen people hostage. It was common knowledge in North Carolina then (and for many today) that drug trafficking was big business in Robeson, a rural county half the size of Rhode Island that is also the midpoint of Interstate 95 between New York City and Miami.[5]

Mab Segrest, now Fuller-Maathai Professor Emeritus of gender and women's studies at Connecticut College, captures well the difficulties and complexities of Robeson County in *Memoir of a Race Traitor* (1994), a narrative that recounts her successes and struggles leading North Carolinians Against Racist and Religious Violence (NCARRV) in the 1980s amid a statewide resurgence of the Ku Klux Klan. In her chapter "Robeson, Bloody Robeson," she describes four murders of particular significance. Joyce Sinclair, thirty-five, a black woman, was raped and stabbed to death in 1985, her body dumped in a field that the year before had been the site of a Klan rally. News reports of her murder stated that she "had just been promoted to a supervisory job in the plant where she worked," Segrest remembers, "a job formerly reserved for white people." Segrest returned to Robeson County in 1986, after the killing of Jimmy Earl Cummings by a sheriff's deputy, narcotics agent, and son of then-sheriff Hubert Stone. Segrest describes a rushed coroner's inquest that ruled that Cummings's death was "an accident or in self-defense," an example of the corruption of the county's legal system that Maurice Geiger, founder of the Rural Justice Center, described three years earlier: "In all of my experience in criminal justice I have never seen a jurisdiction in the United States with such a consistent disregard for fundamental due process, a prosecutor's office so pervasive in its abusive practices, or a judicial attitude that so condones those practices." In the wake of Cummings's killing, Robeson County residents—primarily poor, African American, Lumbee, and white alike—launched Concerned Citizens for Better Government, and newspaper reports suggested that Cummings's death was connected to the theft of drugs from a sheriff's office locker. In 1988, an African American man, Billy McKellar, died in the Robeson County jail because jailers refused him his asthma medication, Segrest recalls, a death that figured in Hatcher and Jacobs's takeover of *The Robesonian* about three months later.[6]

Yet Segrest and NCARRV had to step away following the murder of Julian Pierce, a candidate for a Superior Court judgeship vying against longtime district attorney Joe Freeman Britt, a man hailed by the *Guinness Book of World Records* as America's deadliest prosecutor for the number of defendants he sent to death row. The sheriff's department labeled Pierce's murder a case of domestic violence, committed by a man who subsequently shot himself. But many saw—and continue to see—the killing as an assassination, the ruling power structure's elimination of a man who had helped merge the county's five racially segregated school systems, an attorney who promised reform. In the wake of Pierce's murder, Segrest was "fried," she writes:

> I had become a woman haunted by the dead. . . . I had driven too many nights down the snaking interstate, black and massive around the feeble blaze of headlight, or following spidery lines on maps, beside me on the seat coffee or [C]oke, chocolate or ham biscuits, feeding fear with poison.

Fearful for her safety, haunted by the specter of being followed, suffering from a variety of ailments, Segrest's work in Robeson County and her tenure as director of NCARRV were unsustainable. Even trying to explain to friends her work and its challenges left her ravaged: "I tried to tell [a friend] about Robeson County: *cinderblock buildings, blood in sandy soil, asthma, jail houses, twenty hostages*," she recalls. "I felt incoherent." In the end, Segrest writes, "we went up against too much money and too much death, and we lost big." And those who remained, such as community activist Mac Legerton, changed their approach: "Zealots are necessary for movements, but there's a time when, if you want to stay alive, you have to move beyond that," he tells Segrest. "Now, I'm more into learning than achieving."[7]

Although Segrest writes about events that took place less than thirty years ago, few of our students from outside Robeson County—and only a few from Robeson as well—know anything about this history of institutional corruption, racial violence, and regional activism. Inequities for, and discrimination against, Robeson County's poor residents of all races figure in "Swamp Posse" through specific characters' jobs (or lack of work), experiences with violence, and limited education. Characters like Sheila Ann, Seven, and Marsh illustrate and make personal the myriad ways in which Robeson County continues to grapple with intergenerational problems of health, violence, education, and persistent poverty.

Our first ten years at UNCP have been eye-opening in layering understandings of where we are onto understandings of where we come from. Our work with students, in particular through service learning and community work, has made the litany of problems personal and direct—and has fostered within us an ethical imperative to address these through our teaching. These issues include the facts that

- Robeson County ranks last of North Carolina's one hundred counties in health outcomes: people die earlier, are more likely to suffer obesity and diabetes, and are less likely to have health insurance and easy access to doctors, dentists, and mental health providers, according to the Robert Wood Johnson Foundation.
- Rates of violent crime in Robeson County—at 859 instances of violent crime (murder, rape, robbery, and aggravated assault) per 100,000 people—are among the highest in the state.
- Robeson County spends about $1,000 per student less than the state average, putting it at the bottom of the state's one hundred counties in contribution to K–12 education. Educational outcomes are sobering: From third to eighth grade, students enrolled in the Public Schools of Robeson County score approximately 20 percentage points lower than the state average on end-of-grade tests, with only 20 percent of fifth graders proficient in reading, for example, while only 19 percent of eighth graders are proficient in math, according to the state Department of Public Instruction.
- With a median household income on average of $30,581 (compared to the national average of $53,482), an overall poverty rate of 33 percent, a child poverty rate of 47 percent, and a deep poverty rate of about 15 percent, Robeson County remains a persistently impoverished county, among the most impoverished in the United States.[8]

For too many residents, our students and coworkers alike, Robeson County is no easy place to live.

Having learned these facts over ten years about the region in which we teach, both of us now recognize how much we came to this region with different sets of misconceptions and preconceptions, all of which became even more complicated by the Great Recession that has unfolded since 2008. Both of us clung to a vision of larger places and spaces—the Triangle and Nashville, San Francisco and Los Angeles, sprawling Research I campuses—as the sorts of environments that felt familiar, the types of cultures we understood and that understood us, places that have weathered the economic downturn and even excelled. How could southeastern North Carolina sustain us as emotional and professional beings, especially as a tanked economy has exacerbated our region's devastation through closed businesses, rising unemployment and poverty, and a shredded social safety net? How indeed did this region—its political, economic, human, and more-than-human ecologies—sustain itself, especially in the ways that pervasive historical traumas and ensuing racism, colonialism, sexism, inequality, and homophobia pervert shared consensus and political responses to environmental threats? The relationship between the land and the people here gradually revealed to us answers to some of these questions, exposing our initial ignorance (as well as some kernels of truth) about the place we now inhabit. Locklear's "Swamp Posse" reflects this complicated, enduring relationship between the people and the landscape of Robeson County; it reflects what we have learned about the ways in which these relationships work to sustain themselves in fluid ways, and it nags us about what we have yet to learn here and possibly never will.

"Swamp Posse": The Spell Has Not Worn Off

"Swamp Posse" opens in a moment of insomnia, making clear from the outset that Vince, although returning to his homeland, is no longer fully at ease here. Vince awakens at 12:30 a.m. to what he thinks are the growls of wandering cats outside the mobile home's walls. Besides the cats, the trailer's other palpable noises hold Vince's attention: "the toilet made noises" and "the air conditioner sang" as Vince "felt [his] ear twitch from the sound" of "the angry screech of a cat."[9] The threat of violence lurks here, in a place that is full of life yet far from tranquil.

Vince returns to bed, gently holding his sleeping girlfriend, Sabrina, as she seems likewise discomfited, in the throes of a bad dream. Still, sleep does not come. He gets up again, this time to listen to an oral history he has recorded as part of his graduate coursework; soon, he hears the distant roar of a Harley Davidson. The motorcycle comes closer, and Vince stumbles to the door, knowing that it can only be his friend Harley Oxendine, who arrives with an invitation to join him and their old high school friends for some beer and bullshitting along the banks of the nearby Lumbee River. Vince hesitates about leaving Sabrina only briefly before settling onto the back of Harley's bike and holding on for the short trip to the river. Their friends—Sheila Ann (Vince's high school crush), Seven Jones ("One of the darkest Lumbees [Vince] knew"), and Marsh (a cook at Waffle House)—are already there, drinking rum and beer, a fire blazing on the banks of the river.[10]

In foregrounding Vince's identity as a graduate student, the story makes education a question of personal and communal sustainability. Even if they come from as close by as a crossroads town down the road, leaving home for college, returning to school for retraining or after military service, and imagining the prospect of leaving home after graduation for a job are life-altering considerations for our students. For many students in Robeson County, especially those who are the first or the only in their families to go to college, or who have been able to go to college when siblings, cousins, other relatives, or high school friends have not, attending university can feel like a move away from "home." The distances that can emerge between family members with different levels of education are not always measured in miles, but in perceptions of divergent intellectual, experiential, and emotional opportunities and futures. While families are consistently proud of their UNCP graduates, our students nevertheless often speak of the ways in which family members—sometimes subtly, sometimes overtly—will point out new divisions between themselves and their college-educated relatives, a situation Locklear's story represents through the dynamic of Vince and his crew by the river, which is at once familiar and strained, a dynamic about which he is hypersensitive. For many of our students, then, higher education becomes a way of ensuring a more sustainable economic future, even as it strains the sustainability of specific family relationships.

At the same time, identifying Vince's graduate research as oral history work he is conducting by interviewing his Lumbee relatives privileges local Lumbee cultural knowledge, making it clear that despite Vince's unspoken ambivalence about the ability of Robeson County to provide a viable future for him (and for Sabrina too, perhaps), he nonetheless believes his elders' knowledge should be heard, shared, and preserved, especially as changing times threaten the viability of the places and cultures that comprise Vince's identity. Yet the question remains: Can Robeson County support him and a family for the rest of his life? Or will he, like so many others, take his education and commitment to other places—places that have jobs that fit his training, schools that will care for his future children, neighborhoods that are safer?

Together down by the Lumbee River once again, Vince and his old pals catch up on each other's lives, with Marsh and Harley occasionally ribbing Vince for what they see as his university-educated, white affectations. When Vince refers to his master's research, Marsh jabs, "Oh, listen to that, will ya? Master's degree. Oral history. Research. He sounds like a professor." Seven chimes in by adding, "'Now what the hell is oral history anyway? Is that when you keep up with all the blowjobs you've ever had?" These comments irk Vince, but he works not to show it by laughing along with the group. As Seven becomes increasingly drunk, he wanders close to the river while the others warn him to be careful. They talk about totens, including

Vince's memory of walking through a cornfield with his sister and his Uncle Jake, hearing the rustling of the cornstalks behind them—evidence, Uncle Jake tells the two children, of their being accompanied by the spirit of their Uncle Noel, who was accidentally shot as a child by older brother Jake and has remained with Uncle Jake since then. The inspirited landscape of Robeson County as Locklear represents it pulses with the spirits of those who have gone on before.[11]

The characters' discussion of totens leads to Seven's memory of his deceased brother, David, who he insists was not the victim of suicide, as law enforcement supposed, but was murdered by a white man named Mike. "They said his gun went off and he shot himself," Sheila Ann explains, "but everybody thinks that somebody shot him." Drunk and angered by this memory, Seven vows, "I'm going to shoot somebody one day. I'm going to kill me a nigger or a white man." When Sheila Ann tells him to calm down, he threatens, "I'll kill you too, you whore," pointing at Sheila Ann while telling the group that she "done and fucked half the county. How's it feel to have syphilis, huh?" Sheila Ann retreats into wounded silence.[12]

This section of "Swamp Posse" invites discussion around several important ecocritical themes. Seven's violent language toward African Americans and women allows us to explore with students whether it is possible to achieve ecological sustainability in a social environment where overt racism and sexism persist. During our years at UNCP, we have heard stories from students in which a female student was physically and/or sexually assaulted because "she was asking for it"; likewise, we have heard of a Lumbee student's older relative's refusal to allow the student to bring an African American friend into the house through the front door. Our students have varying responses—repudiation, indignation, resignation, acceptance, agreement—to these events when they relate these stories, but surprise is rarely one of their responses. In 2010, someone spray-painted "Nigger Central" on a billboard off the highway leading into Pembroke, perhaps a reaction of racial anxiety by someone hostile to the perception of an increasing demographic of African American students at UNCP. The perpetrator was never found, and the incident was set aside in time, the billboard replastered. Although such incidents are infrequent, and members of the campus community (such as the Office of Diversity and Inclusion and the Faculty Senate) respond quickly and forcefully to them, the events are nevertheless reminders of an enduring and pernicious layer of the sexism and racism—horizontal and vertical—whose roots in the colonial history of this place are highlighted in the perspectives and actions of the story's characters. Can there ever be authentic environmental justice, we ask our students, without social justice? If all individuals, no matter their race, sex, or gender, cannot gather equally at the table, be heard, and be counted, solutions for the common good cannot take root. And if shared solutions to shared problems are not possible, then the alternative—every man or woman for him- or herself—continues unabated, with the same results of ongoing inequality and injustice.

What's more, an investment in shared solutions for a shared future dissolves in the face of short-term considerations of one's health and well-being. Seven's verbal abuse and misogynistic sexual insults toward Sheila Ann specifically call attention to poor health outcomes prevalent in Robeson County, the result of a lack of consistent, visible access to quality health care in general and to sexual health education and resources in particular. North Carolina's "2013 State Health Profile" reports that 47.3 percent of high school students reported having had sexual intercourse, with 39.2 percent of those who are sexually active not using a condom during their last sexual encounter. As of 2012, North Carolina ranked twenty-fourth in the United States in its rates of primary and secondary syphilis.[13] These realities lie beneath Seven's cruel attack on Sheila Ann as a woman and as a sexual being—an attack that goes unchallenged by the group—as he simultaneously passes moral judgment on her sexual behavior and identifies her as diseased in front of Vince, who was once infatuated with her. This painful scene adds new dimensions to the ambivalence readers sense from Vince about his return to Robeson County, not only temporarily as a graduate-student researcher, but even more so as seeking a sustainable physical and social environment to which to return permanently.

Queer bodies are also the targets of both verbal and physical violence in "Swamp Posse." Such targeting speaks to the ways our region's conservative Christianity and politics continue to make it difficult for LGBTQ people in the campus and larger community to feel safe and supported in openly expressing their sexual identities. Before Seven verbally attacks Sheila Ann, the posse discusses another friend, Tony Ray, who "used to tell some crazy toten stories." When Vince asks what has become of Tony Ray, Seven answers, "You mean [to] tell me you didn't hear about that faggot? He turned queer and moved up [to] Baltimore with some other fag. It was about this time last year that he got killed."[14] Queerness in "Swamp Posse" is represented as both aberrant and unsustainable as a way of living in the world. From Seven's perspective, a perspective not contradicted in the story by any of his friends, to be an openly LGBTQ person is to set oneself up to be the target of justified violence, an ironic attitude in light of Seven's expression of anguish over his own brother's ostensible murder. The fictional Tony Ray, like some LGBTQ students we have had in our classes at UNCP, chose to move away from Robeson County, for reasons the story does not make clear but that readers might interpret to be partly because urban environments (for example, Baltimore, a city with a sizable diasporic community of Lumbee) offer larger and more welcoming communities for LGBTQ people than do rural Southern ones. While we have seen a more inclusive culture develop on campus for LGBTQ students in our time here, we have also heard stories of a gay Lumbee student being beaten up at a gas station in broad daylight and a gay white student who graduated and later committed suicide. We have driven by church marquees in Pembroke reminding congregations that "Marriage is One Man and One Woman" during the campaign for North Carolina's Same-Sex Marriage Amendment of 2012; four years later, we saw similar fears unleashed by the state's passage of House Bill 2, insisting that individuals use public restrooms based on the sex listed on their birth certificates. "Swamp Posse" thus hints at some of the many attitudes and daily realities that manifest in the place where our students' lives intersect with our own.

Back by the banks of the Lumbee, Seven's agitation culminates in his running wildly off into the swamp's wooded darkness. When no one follows him, Vince "[takes] off running beside the river" to try to help him. Vince is eventually able to soothe his friend by telling him a story of how the Lumbee have always survived in the swamps:

> Back in the day, the Lumbee knew spells to make them invisible. They changed and became like swampy mists, like smoke so the white people would not relocate them. And they kept their lands, and adapted and changed, but now, many years later, the spell has not worn off, and we are still an invisible people, like mist.

Seven grows quieter as Vince finishes this story, eventually allowing Vince to pull him to his feet and lead him back to the others by the fire, where the swamp posse's party is getting ready to break up for the night.[15] It is significant that Vince succeeds in calming Seven through a moment of Lumbee oral history, spontaneously and directly applying the focus of his academic research to a personal situation in a way that offers help to a broken friend. In this moment, Locklear emphasizes that despite historical hardships and contemporary challenges, Lumbee culture, living in Lumbee lands, can save lives by sustaining and preserving Lumbee people in the places and communities they call home.

As morning draws near and Seven and Marsh go their separate ways, Sheila Ann stops Vince to tell him that she has done some oral history work of her own. "Yeah, uhm, right before my mamma died, I got me a tape recorder and let her talk into it. I wanted my little girl to have something to remember her by, you know? And so, mamma just started telling these stories, ones that my grandmother told her, stories that my great-grandparents told her." She invites him to listen to her recordings, hopeful they can help him with his research: "'Well, just call me sometime . . . look me up in the phonebook. I'm at home every day. Don't never go nowhere. You can come over whenever you'd like,' she smiled." Vince promises to take her up on her offer and hops on the back of Harley's motorcycle for the ride back to his trailer. Drunkenly ripping off his shirt as they careen through the back roads, Vince "screamed and screamed. It felt so good to scream at the moon. I screamed for what was lost and could never be found again, and I screamed and I screamed 'This is me.' And I screamed, 'I can still remember,' and I screamed 'Kill me!' and I shouted, and I loved and I hated all that was alive, until I couldn't remember my true name anymore."[16] To sail through the nighttime woods, to reunite with old friends, to reconnect with the river: these experiences, these rituals, attach Vince back to his home, even as they remind him of why he left and of things he would like to forget. For Locklear, this kind of deep reconnection drives the story, a way of telling and cementing home and family, family history and ancestors, and community.

Ecocriticism as Sustenance, Preservation, and Perseverance in a Challenging Region

In order to make sense of the signs, and beyond introducing students to local environments through narrative, "Swamp Posse" propels students to consider the larger concepts of ecocriticism that frame the class's discussions of texts and films for the rest of the semester. After a general discussion of plot, character, setting, and other literary elements of the story, we ask student groups to generate ecocritical questions the text raises, and then, as a class, we discuss their brainstorming. The two of us help shape their various questions, many of which overlap, into some overarching themes that illuminate and expand Locklear's story. Foreshadowing the concepts and approaches broached in Garrard's *Ecocriticism*, the students' next required reading, we charge students to read and think like ecocritics about

"Swamp Posse": to undertake "the study of literature as if the environment mattered," in the words of David Mazel. We want them to think about the complexities of the pastoral that Garrard explains, an "underlying narrative structure in which the protagonist leaves civilization for an encounter with non-human nature, then returns having experienced epiphany and renewal," complicated by "the sexual coding of wilderness as a virile, heterosexual space." We hope they will see how "Swamp Posse" pushes back against prevailing paradoxes of wilderness as "wholly pure by virtue of its independence from humans," just as we hope they will see how the story resists idealizing "human subject[s] whose most authentic existence is located precisely there." Last, Garrard prepares them to "dwell" in the story, interpreting and analyzing "the long-term imbrication of humans in a landscape of memory, ancestry and death, of ritual, life and work."[17] The insights Garrard offers and the questions students ask press them to look more deeply at the nuances of "place" in its multiple iterations in the story at the same time their queries reveal much about their understandings and relationships to regional environments, their homeplaces, and cultural communities not their own.

Students' initial ecocritical questions seek to understand the story's place, in particular the significance of the Lumbee River. Their questions include "Was the swamp dry or marshy?" "How high was the river?" "What season might it have been if water levels were at this level?" "What were the fish Marsh was fishing?" The range of questions they ask underscores their intimacy with rivers, either the Lumbee in particular or waterways in general, for hunting or fishing, and with swamps, floodplains, and streams as places of congregation, sustenance, and pleasure. As we reflect on these questions, we conceptualize them as having answers that are knowable and quantifiable—answers we might draw from Google or learn from kin, such as rainfall rates and water levels and fish species native to a blackwater river ecosystem. As we note to students, too, a local person might be reluctant to answer some of these questions if she or he wants to protect a particularly sweet fishing hole or hunting ground. Given the story's influence on provoking ecosystemic questions, though, we hope they will follow up and find these kinds of answers for the ecosystem that surrounds Pembroke, or the ecosystem that surrounds the places from which they come.

As they continue to figure out what they want to learn from the story, students begin asking questions that demand discussion from multiple perspectives to accrue and reveal deeper significances. Students' questions of "Why do they meet at the river?" and "What does the river symbolize to the Lumbee community?" prompt us to consider the cultural and social role of the Lumbee River to the people who populate its watershed. As historian Malinda Maynor Lowery (Lumbee) writes, "I knew first and foremost that I was part of a People; that I had a family and that my family connected to other families; and that all of these families lived in a place, what for us was a sacred homeland: the land along the Lumber River in Robeson County." Vince reflects in "Swamp Posse" that the Lumbee River is a place of beauty that makes a "soft patter . . . as she [runs] downstream," and, despite his awareness that the county's "crime rate was soaring," he "felt safe from all harm. There by the river, [he] found a good feeling."[18] For the Lumbee people who live in southeastern North Carolina, the Lumbee River creates connection and belonging, a living being to which no regional Native person does not have some attachment, and about which everyone has a story. By asking why the characters meet at the river and how that river signifies, students cut to the heart of the complex cultural connections to place and homelands that are not only represented in Locklear's fiction, but that inform Indigenous historical presence and contemporary experience in rural southeastern North Carolina.

Inevitably, our conversation turns toward the unanswerable questions that transcend our discussion of "Swamp Posse," questions to which we will return again and again throughout the term. "Does culture develop in place? Is there a relationship?" students ask. "How does their environment affect the way they speak?" Such questions seek to work out personal, social, and natural connections between humans and place, even as we wrestle with the relationship of culture and environmental belonging: "Are Native people prone to staying within their environment?" Another query—"Do our thoughts have environmental consequences?"—forces us to confront the ability of human beings to sustain human and nonhuman life on the planet through imagining new ways of relating to the earth, on one hand, and the hubris of placing all of nonhuman life and planetary geography at the whim of human imagination, on the other hand.

Students think about how who they are is intricately interwoven with where they are: We encourage them to remember a natural feature of their childhood landscape that has informed their identity as a person in their family, in their town, in their community, in their own skin, as the river has informed Vince and his swamp posse. To deepen this learning, we have asked them to undertake environmental oral history projects in which they conduct interviews with local elders that lead them to extend the range of ecocritical questions we have discussed in class. Local community people and students' grandparents, great-uncles and great-aunts, and other kin who are the primary interviewees for this assignment offer rich details and commentary about what it means to live deeply in places. More importantly, they serve as guides in the flesh of sustaining, preserving, and persevering in a challenging region.

Conclusion: Everybody's Down by the River

This vacillation of belonging and exile at the heart of Chad Locklear's "Swamp Posse," coupled with its setting in the woods and along the river and swamps that surround UNCP's campus, make this story the best and worst of texts with which to begin tackling the issues of sustainability that are the focus of our Environmental Literature course. On the one hand, this text allows us to affirm the Lumbee heritage of our institution and local community and encourage students—Lumbee and non-Lumbee alike—to begin reflecting on their home environments in the intimate and particular ways that Vince describes Robeson County, the Lumbee River, and its swamps. On the other hand, this text confronts us with the problems that beset our community: racism, sexism, homophobia, and violence. In these ways, Vince and students alike struggle with the ambivalence of remaining in place and with what is needed—individually and communally—to sustain vitality in individuals, cultures, and natural environments. As professors of our students in this place, we too struggle with having witnessed both the best and the worst of the themes this story expresses in Robeson County and elsewhere in North Carolina; it forces us, as faculty, to think deeply about our own responsibilities to the landscape in which we teach and to the students who call the region home, and about our abilities and capacities to sustain ourselves personally and professionally here.

Ultimately, maybe the real question isn't whether to stay or go at all, but *how* to stay if you stay, how to go when you go, and how to find your way back when and if that time comes. If such thinking helps our students foster sustainable ways of living in places, wherever those may be; of listening to those places' stories; and of creating stories of their own that offer hopeful and realistic possibilities for sustainability, we will be content. "'You art to come with me,' Harley tells Vince in "Swamp Posse." "Everybody's down by the river."[19] Chad Locklear guides us to the river, from which we'll find different ways back to tell the stories of what we've learned there, and how we can keep learning through the land, the water, and the people and places that are rooted here.

NOTES

1. Glenn Ellen Starr Stilling, "Lumbee Indians," in *Encyclopedia of North Carolina*, ed. William S. Powell (Chapel Hill: University of North Carolina Press, 2006), 699–703, http://lumbee.library.appstate.edu/bibliography/stil007; Chad Locklear, "Swamp Posse," *Pembroke Magazine* 38 (2006): 175.
2. Whereas the river was designated the Lumber River by the state of North Carolina in 1809, the American Indians who inhabited the region traditionally have referred to the river as the Lumbee River, a history that Lawrence Locklear describes in "Down by the Ol' Lumbee: An Investigation into the Origin and Use of the Word 'Lumbee' Prior to 1952," *Native South* 3 (2010): 103–117, http://dx.doi.org/10.1353/nso.2010.0004. In this chapter, we use the name of the river as it is known by Lumbee people and as it is labeled in the story.
3. For an overview of the coastal-plain blackwater river and floodplain that is the dominant ecosystem of our region, see David Blevins and Michael P. Schafale, *Wild North Carolina: Discovering the Wonders of Our State's Natural Communities* (Chapel Hill: University of North Carolina Press, 2011); North Carolina Wildlife Resources Commission, "Floodplain Forest: Mid-Atlantic Coastal Plain," http://www.ncwildlife.org/Portals/0/Conserving/documents/Coast/CP_Floodplain_forest.pdf; and North Carolina State Parks, Lumber River State Park, "Ecology," http://www.ncparks.gov/lumber-river-state-park/ecology.
4. Jeremy L. Caradonna, *Sustainability: A History* (New York: Oxford University Press, 2014), 13.
5. Southeastern North Carolina might be seen as a reviled region within a reviled state. In *White Trash: The 400-Year Untold History of Class in America* (New York: Viking, 2016), Nancy Isenberg labels North Carolina as the United States' "first white trash colony." See Dwight Garner, "Review: 'White Trash' Ruminates on an American Underclass," *New York Times*, June 21, 2016; Scott Raab, "Reasonable Doubt," *GQ*, March 1994, 240ff.
6. Mab Segrest, *Memoir of a Race Traitor* (Cambridge, MA: South End Press, 1994), 79, 107–108.
7. Ibid., 126, 127–128, 123, 132.
8. Robert Wood Johnson Foundation and University of Wisconsin Population Health Institute, "Robeson (RO)," *County Health Rankings & Roadmaps*, http://www.countyhealthrankings.org; North Carolina State Bureau of Investigation, *Crime in North Carolina—2014: Annual Summary Report of 2014 Uniform Crime Reporting Data*, November 2015, http://crimereporting.ncsbi.gov/public/2014/ASR/2014%20Annual%20Summary.pdf; Emma Swift Lee and Joe Ableidinger, *2016 Local School Finance Study*, Public School Forum of North Carolina, https://www.ncforum.org; North Carolina Department of Public Instruction, "Public Schools of Robeson County Welcome Letter,"

NC School Report Cards, http://www.ncreportcards.org/src/distDetails.jsp?Page=2&pLEACode=780&pYear=2012-2013&pDataType=1; North Carolina Justice Center, "2010 Poverty and Deep Poverty Estimates," http://pulse.ncpolicywatch.org/wp-content/uploads/2012/03/NC-County-Data-Poverty-and-Deep-Poverty-Rates.pdf.

9. Locklear, "Swamp Posse," 172.
10. Ibid., 174.
11. Ibid., 175–176, 178–179.
12. Ibid., 179.
13. U.S. Centers for Disease Control, National Center for HIV/AIDS, Viral Hepatitis, STDs, and TB Prevention, "North Carolina—2013 State Health Profile," http://www.cdc.gov/nchhstp/stateprofiles/pdf/North_Carolina_profile.pdf.
14. Locklear, "Swamp Posse," 179.
15. Ibid., 180.
16. Ibid., 181.
17. David Mazel, ed., *A Century of Early Ecocriticism* (Athens: University of Georgia Press, 2001), 1; Greg Garrard, *Ecocriticism*, 2nd ed. (New York: Routledge, 2012), 54, 60, 78, 117.
18. Malinda Maynor Lowery, *Lumbee Indians in the Jim Crow South: Race, Identity, and the Making of a Nation* (Chapel Hill: University of North Carolina Press, 2010), xiii; Locklear, "Swamp Posse," 175–176.
19. Locklear, "Swamp Posse," 173.

BIBLIOGRAPHY

Blevins, David, and Michael P. Schafale. *Wild North Carolina: Discovering the Wonders of Our State's Natural Communities*. Chapel Hill: University of North Carolina Press, 2011.

Caradonna, Jeremy L. *Sustainability: A History*. New York: Oxford University Press, 2014.

Garrard, Greg. *Ecocriticism*. 2nd ed. New York: Routledge, 2012.

Locklear, Chad. "Swamp Posse." *Pembroke Magazine* 38 (2006): 172–181.

Locklear, Lawrence. "Down by the Ol' Lumbee: An Investigation into the Origin and Use of the Word 'Lumbee' Prior to 1952." *Native South* 3 (2010): 103–117. Http://dx.doi.org/10.1353/nso.2010.0004.

Lowery, Malinda Maynor. *Lumbee Indians in the Jim Crow South: Race, Identity, and the Making of a Nation*. Chapel Hill: University of North Carolina Press, 2010.

Mazel, David, ed. *A Century of Early Ecocriticism*. Athens: University of Georgia Press, 2001.

Segrest, Mab. *Memoir of a Race Traitor*. Cambridge, MA: South End Press, 1994.

Getting Your Feet Wet: Teaching Climate Change at Ground Zero

Daniel Spoth

> Where you come from is gone, where you thought you were going to never was there, and where you are is no good unless you can get away from it. Where is there a place for you to be? No place.
>
> —Flannery O'Connor, *Wise Blood*

On a muggy day in May 2013, the environmentalist, author, and activist Bill McKibben took the stage at Eckerd College's fiftieth commencement in St. Petersburg, Florida. The graduation tent stood no more than a hundred yards across a manicured lawn from Eckerd's South Beach and the adjacent waters of Boca Ciega Bay, where sailboats on the college's waterfront drifted serenely by. McKibben, one of the nation's foremost climate change activists, didn't miss the opportunity to comment on the vulnerability of the backdrop: "We're here at the fiftieth Eckerd commencement," he said. "If we don't get it right, the 100th commencement won't be right here because this will be underwater."[1] Though some of the students (and many of the parents) in the audience seemed to evince a mild horror at McKibben's deviation from the typical tenor of the commencement address (valedictions, futurity, stresses on hope and promise), none seemed taken by surprise. The mood, and the applause that followed, was somber and restrained.

McKibben was followed by John Lasseter, chief creative officer of Disney Animation Studios, who delivered a heartwarming speech to thunderous acclaim about the necessity and rewards of following one's dreams.

As I walked away from the graduation tent that day, my black academic robes feeling as hot as a cast-iron skillet in the Florida sun, I ruminated on the juxtaposition of the two speakers, on the precarious position that we (we everywhere, but we on this college campus in particular) find ourselves occupying, and on the challenges and ironies of teaching environmental topics against such a backdrop. I teach environmental literature on a waterfront campus whose altitude doesn't reach above seven feet, where days of heavy rain close parking lots, and faculty write contingency clauses for catastrophic hurricane damage into their syllabi. Teaching about climate change and ocean level rise in this environment becomes, for me and my colleagues, not a set of vague predictions and projections, but a very real and present issue, one that will, with certainty, change the way that we live and work in this place in profound ways. But this essay is not meant to be a catalog of the environmental threats that face the college so much as an investigation of what can be done on a pedagogical level to ensure that our graduates are the sort of leaders who can address these threats and the obstacles that stand in the way of that goal. As I will argue presently, the unique challenges that sustainability education faces in the type of environment in which I teach fall roughly into three categories: the necessity of cultivating a sense of place, the difficulty of conceiving of worldwide threats, and the problem of navigating the divide between hope and despair.

Certainly there are many *practical* actions that universities can take in order to make themselves and the world that surrounds them more sustainable—movements toward divestment of endowments in fossil fuel interests (a stratagem that McKibben has heavily promoted); environmentally focused service-learning projects; on-campus gardens, henhouses, composting and recycling stations; green and carbon-neutral buildings—all of these are evidence of colleges acting, and not merely thinking, progressively. And all of the above have either been successfully implemented or are in the planning stages of implementation at Eckerd. David W. Orr is likely correct when he states, in *Down to the Wire* (2009), that "were [universities] to . . . use their buying and investment power to build local and regional resilience, they could greatly speed the transition to a decent future."[2] Yet, ultimately, I am more concerned with the avowed pedagogical mission of higher education than how college resources might be leveraged toward more sustainable ends; as I shall discuss later, our potential ideological contribution far outweighs the material contribution that we can make toward sustainability.

In this regard, we as educators seem to be doing a somewhat poorer job of addressing sustainability than we could be. The consensus reached by scholars surveying the efficacy of educating for sustainability is that we require more and better efforts on the part of faculty and academic affairs administrators. David Sobel writes, in *Place-Based Education* (2004), that "schools and other educational institutions can and should play a central role in [the process of healing nature and community], but for the most part they do not." Similarly, C. A. Bowers writes in *Educating for an Ecologically Sustainable Culture* (1995) that often colleges foster destructive rather than sustainable attitudes in their graduates: "Most citizens appear not to recognize the connections between the Western ideas and values they were inculcated with in schools and universities, their consumer-oriented life style, and the depletion of fish stocks, aquifers, old growth forests, petroleum reserves, and the accumulation of toxic wastes at all levels of the biosphere." To be sure, higher education has become considerably more environmentally focused in the two decades since Bowers made the above claim, particularly in small liberal arts colleges such as Eckerd. Yet I frequently find it necessary to drive home even basic lines of causality between the late-capitalist ideals that my students take for granted and the presence of looming environmental catastrophe. Regardless of any current shortcomings in the field of sustainability, however, my colleagues and I hold fast to the belief that college education has the *potential* to transform our students' environmental ethics. While a 2005 United Nations report claimed that "at current levels of unsustainable practice and overconsumption it could be concluded that education is part of the problem," it also suggested that "a deeper critique and a broader vision for the future" could make education part of the solution as well.[3]

Though scholarship on sustainability pedagogy frequently stresses community engagement and service learning, I believe that the campus itself also offers massive potential for promoting sustainable ideologies and practices as well as cultivating a sense of, and love for, a particular place, a prerequisite for sustainable thinking writ large. Too often the physical presence of the college campus is elided or dismissed as a simple backdrop against which the real "work" of academia is done. Mark S. Cladis bemoans the fact that students are frequently pushed to regard the intellectual aspects of their college education as entirely independent of the more solid, tangible elements of living and abiding in a particular place: "We wish to protect them from distractions, from drudgery, from alleged impediments to intellectual growth. In so doing, we treat our students as Disembodied Minds. We arrange everything so they can read, write, and think with great efficiency, unburdened by the tasks of Embodied Living." It is, however, possible to integrate learning within the classroom with the campus environment outside the classroom; multiple campus garden programs (including Eckerd's) are joined with traditional academic courses, for example, providing opportunities for students to learn in a variety of settings without ever setting foot off campus. Suzanne Savanick Hansen, in fact, marks the use of the campus as a teaching tool as "part of a liberal education, the kind of education we need if our society is to make the changes needed for a sustainable future."[4]

Though Eckerd's campus offers significant resources for teaching sustainability by simple virtue of the location and nature of that campus, I would like to speak first about the distinct obstacles imposed by our particular place. On one hand, our campus is located in coastal Florida, one of the areas most likely to be first affected by sea-level rise, in a state whose politicians are steadfastly devoted to denying its existence. Current two-term governor Rick Scott, who in early 2015 banned the phrases "global warming," "climate change," and "sustainability" from any official statements or communications, and whose gubernatorial career has been termed an "environmental disaster" by the *Tampa Bay Times*, continues to ensure that the state will take no large-scale political action either to alleviate the progress of climate change or to prepare for its effects. In August 2014, five climate scientists from around Florida—including Eckerd's David Hastings, professor of marine science and chemistry—convened with Governor Scott in a desperate attempt to convince him of the necessity of taking action in response to climate change, or, at the very least, admitting that climate change was a very real threat to the state.[5] This meeting, perhaps unsurprisingly, resulted in no new policy or even rhetorical shifts on the part of the state government. In short, the scientific—and, to a large extent, the educated—population of the state's advocacy for climate change legislation and action has been repeatedly stifled by regressive politics.

But our student body is not primarily Floridian; the majority of Eckerd students are from the mid-Atlantic states, the Midwest, or New England, representing a total of forty-eight states and forty countries. Fewer than 25 percent come from Florida itself, and the bulk of the student population is, by and large, unfamiliar with the long history of climate-change advocacy and conflict within the state. And herein lies one of the primary challenges in dealing with environmental topics in our classes: most of our students' experience with the state is by nature fleeting, temporary, even casual. Frequently, Eckerd students are raised in other states, attend college in Florida, and, after graduation, return to their hometowns or depart for other locations across the world; our student population is extremely cosmopolitan. Like the infamous "snowbirds," wealthy residents of colder climates who winter in Florida, Eckerd students frequently spend academic semesters at the college while still regarding themselves as residents of—or, more bluntly, "from"—other states. While this diversity of origins certainly creates a lively student body, it also signifies a certain environmental transience; there is a distinct feeling that though the faculty strive to communicate the necessity of sustainable practice to their students, most of those same students view their college environment as distinctly temporary, even disposable, a train of thought that has been carried out from early twentieth-century accounts of agricultural opportunism to contemporary media such as *Miami Vice* and *Spring Breakers*. Florida has been regarded as a "winter state," or place of distinctly temporary leisure, since its inception. Though a complete account of this mode of representation is beyond the scope of this essay, suffice it to say that the very location of the Eckerd campus acts to counteract the cultivation of the most basic of all environmental precepts: a sense of place.

Cultivating a Sense of Place

"A sense of place" has long been a byword of ecocriticism and regionalist literature. As a means of approaching the concept, we might do worse than Eudora Welty's seminal essay "Place in Fiction" (1955), which marks a sense of place as being "as essential to good and honest writing as a logical mind." Welty succeeded in putting into words a general sentiment toward the concept of "place" that had occupied writers for decades before her lecture saw print, and it is still cited today as evidential of the power of place in media and contemporary thought. Welty contends that representations of place contribute three separate elements to good fiction writing: a material backdrop against which the events of the story can proceed, a particular worldview or lens through which the actions of characters can be seen and colored, and a sense of identity for the writer. Of these, the third has proven both the most essential and most nebulous principle, since it is deeply subjective and sentimental. Welty writes: "It is by the nature of fiction itself that fiction is all bound up in the local. The internal reason for that is surely that *feelings* are bound up in place." For Welty, as for most nature writers, these sentiments form a pattern of wonder, reverence, attachment, and even spirituality, and her own comments on the value of place in fiction take on those same tones: "From the dawn of man's imagination, place has enshrined the spirit; as soon as man stopped wandering and stood still and looked about him, he found a god in that place; and from then on, that was where the god abided and spoke from if ever he spoke."[6] Here, an appreciation of a place appears as something entirely distinct from any aesthetic pleasure derived from the landscape or a kinship with the manners and traditions of its people. Rather, Welty's sense of place is inextricable from notions of stasis and duration in a particular place, what Heidegger would call "dwelling"—living syntagmatically and purposefully.

Though situating place at the epicenter of a writerly consciousness was relatively unproblematic for Welty, academia for roughly the last half-century has tended to emphasize "space" over "place," a transformation that Anthony Giddens, in *The Consequences of Modernity* (1990), attributes to the interpenetration of local and distant regions enabled by modern technologies of communication and transportation. The "spatial turn," as it is sometimes called, signifies a defetishization of individual regions in favor of a more egalitarian, permeable, and mobile configuration. Rather than the subjective attachments and auras that animated Welty's view of place, "space" offers a homogenization of individual places into networks of social and structural forces—the public sphere, for instance, or the African diaspora. Barbara Ladd characterizes this turn as "do[ing] away with questions of situatedness and with questions of relative value" that might lead criticism toward essentialism and inertia. Indeed, paeans to the enduring value of place today frequently come across as old-fashioned and needlessly reductive, distinctly unwelcome amidst contemporary theories of globalization. In the introduction to *Cosmopolitics* (1998), for example, Bruce Robbins finds the concept of place to be a cold comfort in an increasingly connected world:

> The devastation covered over by complacent talk of globalization is of course very real. But precisely *because* it is real, we cannot be content to set against it only the childish reassurance of belonging to "a" place . . . yes, we are connected to the earth—but not "a" place on it . . . we are connected to all sorts of places, causally if not always consciously, including many that we have never traveled to, that we have perhaps only seen on television.

Here, Robbins offers an alternative conception of place that is very much at odds with Welty's earlier definition; rather than signifying stability and individual attachment, place persists in the spatial zeitgeist as an unmoored, locationless concept, or as a media creation. This is, in essence, the endpoint of Edward S. Casey's prediction, in *The Fate of Place* (1997), that "the gradual and forceful encroachment of space upon place" would conclude in "the virtual disappearance of the latter into the former."[7]

Yet even as some scholars proclaim the death of the concept (or at least the utility) of place, others continue to insist upon its relevance and value. Tim Cresswell claims that "place is the very bedrock of our humanity. . . . It cannot have vanished because it is a necessary part of the human condition." Along the same lines, Arturo Escobar suggests that place, which he takes to mean "some measure of groundedness, sense of boundaries, and connection to everyday life," continues, despite its unpopularity in the academic community, "to be important in the lives of many people" regardless of the fact that "its identity is constructed and never fixed." Perhaps the most equivocal perspective on the space/place divide advocates for retaining both terms, regarding them as mutually constitutive and both necessary in order to create a sense of balance in human habitation. To simplify this argument somewhat, subjects exist and move in space, but require specific spaces that are saturated with personal meaning—places—in order to maintain a respect for environments and their inhabitants. Yi-Fu Tuan writes, in *Space and Place* (1977), that "human beings require both space and place. Human lives are a dialectical movement between shelter and venture, attachment and freedom."[8]

In my own pedagogy, I tend toward Tuan's and Welty's somewhat more old-fashioned, rooted conception of place rather than the spatialized cosmopolitanism of Robbins and Casey. The texts I assign, from John Muir and Charles Chesnutt's nineteenth-century accounts of the environment and ethnography of the South to the more modern regionalist writings of William Faulkner, Zora Neale Hurston, Janisse Ray, and Welty herself, all attempt, I tell my students, to capture something essential and vital about a particular location on earth. I accompany these readings with queries that, though they may seem academic in nature, are neither idle nor inconsequential: What does a deep attachment to place look like in literature? What is the artistic value of such an attachment? What can we, as migratory subjects in a connected world, learn from these writings? The evaluative turn of these questions reflects my students' own tendency toward cosmopolitanism—in reality, many of them likely identify with Robbins's claim that we can be "connected to all sorts of places," including many with which we have no direct experience. It may be worthwhile to ask these same questions of my own teaching enterprise—why insist upon the importance of place-based writing when the students receiving that lesson live in a radically different world and under radically different circumstances than those authors?

Pedagogical scholarship marks the cultivation of a sense of place as essential for developing a curriculum of sustainability. Sharon M. Meagher states directly that "students cannot grasp the concept of sustainability unless they recognize the importance of place." Why is a sense of place a prerequisite for learning sustainably? Some scholars, such as Frederick O. Waage, cite a host of benefits to students who are taught a sense of place, including "sensitivity and gentleness toward the land, acceptance of an order of creation not dominated by our own imagery, and respect for other creatures as ethical equals," while David Gruenewald adds "a concrete focus for cultural study" and an expanded cultural scope that adds "related ecosystems, bioregions, and the place-specific interactions between the human and the more-than-human world" to the list.[9] However, I believe that, apart from expanding the scope of sustainability education to include new topics in ethics, philosophy, and the sciences, place-based pedagogies, on a very basic level, encourage a deep emotional and aesthetic *investment* in particular locations on the part of students; acknowledging and connecting with a place is the first step toward encouraging our students to regard that place as worth their respect, perhaps even worth the effort of preservation and restoration.

What, in concrete terms, does place-based education entail? At its most conservative, the term denotes a simple curricular and interdisciplinary focus on a particular region—a traditional course based on the ecology, geography, literature, history, cultural diversity, politics, and so forth, of California, for instance, or the southeastern United States. However, in its broader application, place-based learning strives to overcome what Gruenewald and Smith call "the traditional isolation of schooling from community life," making education "a larger community effort." The recent popularity of universal service learning and community outreach in many American universities is, in some measure, an effort to create a sense of place. However, cultivating the authentic sense of place that Welty and others are interested in requires significant dedication and investment of time, resources, and faculty effort. One such well-supported program is the Center for Place-Based Learning (CPBE) at Antioch University New England in Keene, New Hampshire. CPBE fully integrates academic topics with service-learning projects in order to balance ideology and practice. Steve Chase, a faculty member at Antioch, writes that his students are "passionate about preserving wild nature, improving public health, and creating a sustainable way of life," but are also out of tune with "social oppression, political economy, or the history of people's movements in this country." Chase describes, in his article, a process of both teaching a traditional discussion-based course on environmental issues and linking the topics of class discussion to community-service field trips at the Dudley Street Neighborhood Initiative in a poverty-stricken area of Boston. Essentially, Chase attempts to bring his students to a fuller understanding of the place in which they live by physically placing them in contact with locations and people whom they would not normally encounter, leading to a deeper and more nuanced understanding of that place.[10]

As I have mentioned earlier, at Eckerd we face a somewhat more difficult task in cultivating a sense of place: in contrast to more localized liberal arts colleges or regional universities, where the student body has frequently been born and raised within a few hours' drive of the campus and thus likely feels a strong preexisting attachment to the local environs, the average Eckerd student travels roughly one thousand miles to attend the school. The Admissions office is fond of quoting this statistic owing to the fact that it proves the college to be desirable enough to voyage across half a continent to attend, but our geographical distribution also raises certain pedagogical challenges—independent, of course, of the carbon footprint of so many students traveling such immense distances to attend the school and return home for visits on a regular basis. I have found the bulk of my classes to be filled with extraordinarily dedicated and environmentally conscious students, well aware of the hazards that threaten our campus and the world at large. However, this trend is not universal, and it is also possible to find students who regard Eckerd, or Florida as a whole, with an ethos that is familiar from late nineteenth- and early twentieth-century accounts of the spread of extractive industry, not to mention the Florida land boom of the 1920s and its twin, the real-estate bubble of the early twenty-first century. This ethos might be summarized as this: find an exploitable landscape, derive as much profit from it as possible in as little time as possible, and leave before the consequences set in. This attitude has always been pervasive in our state, though what is extracted today bears less of a resemblance to the gold, oil, and timber of the Industrial Age than the opportunities for cheap real estate, leisure, and entertainment of late capitalism. But the underlying philosophy is the same: take what you can, while it's still there for the taking. As a result, we see on occasion a set of beliefs and lifestyle choices on our students' part that I have come to call "the life of exhaustion," an opportunistic attempt to bleed a place of all that it has left while it's still possible to do so.[11] This ideology is all the more pernicious because it tends to cloak itself in pragmatism—if the state is going to sink into the ocean anyway, that is, we might as well turn up the air conditioning!

Thus, I believe that cultivating a sense of place on our campus is vital to any effort at teaching sustainability, and we as educators must use our own places, our campuses, and the very real challenges that confront them, as tools to that end. This effort is achievable whether the campus is as evidently threatened as Eckerd or not. A liberal arts college in Colorado, for instance, well clear of the advance of seawater, might instead ask students where the food served by its cafeteria comes from, and what might happen to that food source if global temperatures were to rise several degrees, or what might transpire if the snowpack is not replenished sufficiently to fill rivers and reservoirs. Only by causing students to internalize their own involvement and complicity in global environmental issues, to regard their own place (indeed, any place that they come to occupy) as impacted by their own actions and the actions of other, distant subjects, can we significantly raise environmental awareness. In a word, we must instruct students in what Orr calls "the arts of inhabitation" in order to effect "the unlearning of old habits of waste and dependency." Thus far, this "art" has been an understressed aspect of higher education. Gruenewald and Smith mark the elision of place-based education as endemic to not merely college campuses, but the entire twenty-first-century *Weltanschauung*, in which "many people only 'reside' where they live, and develop no particular connection to their human and non-human environments," an attitude that results in "alienation from others and a lack of participation in the social and political life of communities." Mark DiMaggio remarks, perhaps even more dolorously, that college graduates are "increasingly effective at exploiting the resources of the natural world, but continue to operate under the archaic premise that resources are infinite."[12] Encouraging students to connect in a deep and contemplative manner with their place on earth—and, perhaps more importantly, the limits of that place—is, I believe, the first step to overturning these misconceptions.

Typically, my Eckerd humanities colleagues and I approach cultivating a sense of place through an emphasis on students' individual histories and creative interpretations of the landscapes of their past and present. Such an approach is in keeping with the avowed intent of small liberal arts colleges to offer a deeply personalized, sensitive education as an alternative to the more standardized curricula of larger universities. However, contemporary scholarship on sustainability education frequently regards such a focus as inimical to the ultimate goals of environmentalist pedagogy. Bowers performs an extended attack upon "the current emphasis on individually-centered creativity" in favor of "communication between generations, and across species, about how to live in ecologically sustainable relationships." In Bowers's view, focusing on individualism severely limits "development of a sense of identity that incorporates the multiple relationships and memory webs that make up the environment" in favor of an egotistical, anthropocentric perspective. Though Bowers's book represents perhaps the most sustained attack on the individual emphasis in higher education, John Blewitt similarly contends that "thinking and acting sustainably requires individuals to recognize that they are individuals only at the expense of severing connections with the wider social and natural worlds."[13] Thus, as educators, we are in the uncomfortable—not to say untenable—position of encouraging students to develop deeply personal and subjective relationships with equally specific landscapes, yet doing so in a manner that emphasizes community and global relationships over subjective individual impressions.

This paradox, in miniature, is at the heart of a much larger challenge in teaching sustainability in threatened environments. As I have mentioned earlier, for the most part Eckerd students are extremely environmentally aware, motivated, and determined to have an impact upon encroaching threats to world health. By and large, they have worked ceaselessly to raise awareness of climate change both on campus and in the world at large; many of these projects, in fact, are extracurricular rather than incorporated with coursework. Several years ago, a group of students marked all campus buildings with blue tape to indicate the extent of a six-foot rise in sea levels. The cellophane waters drowned the cafeteria, the gym, and many of the academic buildings, and lapped gently at the entrance to the library. This acute awareness of the nearness and immediacy of climate change, perhaps, led Eckerd to send the most students, per capita, of any college in the country to the "Forward on Climate" rally in Washington in February 2013, a fact that McKibben, in his commencement speech, viewed as evidence that "this is a key place for the future, and you are key players for the future."[14] Yet Eckerd students, as well as the faculty who teach and advise them, must come to terms—despite the overwhelming feeling of being at the forefront of climate change—with the uncomfortable truth that their individual agency, even when backed by the resources of the college, is extremely small. Regardless of the moves toward sustainability that we adopt, we will inevitably suffer the consequences of the poor environmental choices of not only the voters and policymakers in our own state, but those of others across the street, across town, and across the world.

Centering Place in the Classroom

I teach several classes oriented toward environmental and sustainability education at Eckerd, among them "Southern Literature and the Environment," which approaches environmental exploitation and natural disasters in the region through my own research specialty in Southern literature, and "Regional American Literature," which explicitly orients itself toward building a sense of place in its constituency by pairing readings on literary approaches to place with service-learning projects conducted at a local nature preserve. A summer term class, "Bombs, Zombies, Plagues, and Waste," also approaches multiple narratives of environmental depletion and collapse. In this course, I frequently run students through a role-playing scenario: the world has been destroyed, I tell them, by climate change, water scarcity, nuclear attack, supervirus, or any number of other calamities. Their task, as a group, is to rebuild civilization using their combined knowledge and skill set. At first, things go well—the group collects food; accumulates fuel, weapons, and shelter; and establishes a thriving small community. My students are, as a rule, excellent at reconstructing society in miniature. But then the scope broadens: more villagers are born, refugees trickle in, survivors of a second, destroyed community arrive seeking shelter. The environment becomes more hostile: harsher winters, wild animals, diseases, and opportunistic competitors leach away their stockpiles. Resources are limited. Crime becomes a problem. When the settlement collapses (as it always does), I discuss what lessons can be gleaned from the experience. We have a fantasy of survivalism on a small scale, my students conclude, but when new threats enlarge our sense of the world, that fantasy dissolves. This is one of the major issues in addressing large environmental problems, I tell them—in cultivating and developing one sense of place, we frequently forget that our place is one among many others.

Such scenarios breed cognitive dissonance among my students—as in the paradox of place, we wish to conceive of ourselves as having (within our own private spheres) some influence on the fate of our individual places, if not the wider globe, yet we are correspondingly driven to acknowledge the fact that these places will be devastatingly impacted by global phenomena well beyond our control. Lynn Mortensen phrases this quandary in terms of the scope of worldwide environmental issues: "The spatial scale of the globe," she writes, "is difficult to grasp and [it is] even more difficult to understand how an individual may influence such a vast entity." Orr calls this intellectual paralysis a "consensus trance," in which the individual remains "oblivious to the full scope, scale, severity, and duration" of the global environmental issues afflicting their particular place. Large, global environmental issues lack simple, straightforward solutions that are achievable through individual actions—in fact, such challenges are difficult even to conceive of. Danielle Lake writes that contemporary environmental threats are difficult to conceptualize because "there is most often no single root problem," and that, ultimately, "multiple factors feed into the problem and . . . this problem is linked to other various communal problems."[15] Conceiving of the degradation of the global ecosystem itself is daunting, and the knowledge that individual agency in regard to these changes is necessarily limited can frequently lead to despair.

My own tendency, when I first began teaching environmental literature and ecocriticism, was to place emphasis on the presence and rapid exacerbation of environmental threats in hopes of creating a sense of urgency and, subsequently, action in my students. And, of course, one does not need to look far in order to uncover examples of environmental degradation so egregious as to utterly horrify such an audience; my own teaching focused on narratives of environmental collapse, from McKibbens's own *Eaarth* (2010) to Octavia Butler's *Parable of the Sower* (1993) and Cormac McCarthy's *The Road* (2006). My students commonly interpreted the preponderance of impending disaster described in these texts in the same manner as Orr: a continuous parade of ill tidings with no clear end in sight.

> The news about climate, oceans, species, and all of the collateral human consequences will get a great deal worse for a long time before it gets better. The reasons for authentic hope are on a farther horizon, centuries ahead when we have managed to stabilize the carbon cycle and reduce carbon levels close to their preindustrial levels, stopped the hemorrhaging of life on Earth, restored the chemical balance of the oceans, and created governments and economies calibrated to the realities of the biosphere and to the diminished ecologies of the postcarbon world.

Elsewhere in the same volume, Orr states bluntly: "I know of no purely rational reason for anyone to be optimistic about the human future." Combined with the factors I have noted earlier in this essay—the lack of a sense of place, the treatment of places as disposable, and the conceptual difficulties in grappling with large global issues—the relentlessly grim outlook on climate change feeds not into shock, determination, and action, as I had originally hoped, but into paralyzing despair. Meaningful change, especially as enacted by a single individual or community, seems beyond the realm of possibility, and what minor victories exist become, as Robert Figueroa writes, "lame compromises of human life and socioenvironmental values against political and economic agendas."[16] Despair is, in itself, not a productive response. While anger, outrage, and even desperation may lead to valuable results, despair produces only inaction and, frequently, the sort of moral and ethical callousness that I have noted earlier.

The solution to these despairing attitudes, according to many scholars, is to temper the more devastating aspects of environmental change with an emphasis on hope and empowerment. Jon Jensen claims that too much stress on the inevitable disasters associated with climate change in a curriculum "backfires in the end and leaves students feeling powerless in the face of certain doom . . . Students need to feel that they have agency and that their work makes a difference, that they can tackle a problem and solve it." Orr, in turn, redirects the conflict somewhat towards a focus on endurance and stoicism in the face of increasingly melancholic developments. "Deep environmental educators," he writes, "must therefore equip students with the stamina to witness ecological losses and collateral social damages without being immobilized by despair. They will need our help to transform their grief into a stronger and deeper attachment to life and a more authentic hope that lies on a farther horizon."[17] Indeed, I find that one of the primary challenges involved in teaching environmental issues at Eckerd involves balancing the necessity of creating a feeling of agency and hope with the inevitability of overwhelming despair in the fact of global threats, made doubly present by the proximity of the campus to those threats. This ability to balance is assuming, of course, that we as educators possess the "stamina" that Jensen alludes to.

Maintaining contemporaneity with new developments in environmental thought, whether in the sciences or the humanities, combined with cycles of class sessions that rework the same issues without a clear view of potential paths to improvement while the news becomes grimmer and grimmer, is a process at least as wearying to the teacher of sustainability as the barrage of apocalyptic predictions that the students themselves endure. For myself, hope lies not in a vision of a reincarnated world in my own lifetime, but in the belief that by displaying to students the changes that are occurring—right in their own backyards, in fact—for them to see with their own eyes, and by marking those changes as part of a longstanding set of attitudes and ethics toward nature that can be decoded through literature and media, I can in effect convince them that something *must* be done, and it must be done by people who have received the sort of education that they are getting.

Thus I have begun, in my own classes, to discuss topics that I had thought to be anathema earlier in my teaching career. How does the class react, emotionally, to learning about climate change? How does reading oral reports from survivors of Hurricane Katrina make them feel? What separates this emotional response from their feelings while walking out of the theater after viewing Benh Zeitlin's *Beasts of the Southern Wild* (2012)? While it may seem needlessly indulgent to devote class time to discussions of the emotional content of environmental disaster, I am finding it increasingly necessary in order to help create graduates who are not paralyzed by indecision and fear. Catherine O'Brien writes, in "Sustainable Happiness and Education" (2012), that while "a focus on happiness could appear to be a diversion from the hard issues of sustainability," such an approach is essential because "there is a natural connection between sustainability and positive psychology." This is not to say that we should manufacture happiness for our students, or offer them false hope, which would perhaps be as bad as no hope at all. Ryan Anderson, in a 2015 essay, counsels instead a balance between hope and "critical awareness," a process of building alternatives that encourages "a more attuned engagement with students which brings to the forefront both others' activism and their own potential action."[18]

Endangered Places as Catalysts for Action

In this essay, I have primarily enumerated challenges to teaching sustainability, the difficulties and paradoxes we face—at Eckerd in particular, but more widely as well. It does seem to be an axiom of environmental education that these challenges will always be more abundant than our advantages, at least for the foreseeable future. Yet I am inclined to believe that at least some of the obstacles that we face as teachers of sustainability may conceivably be repurposed as assets. At Eckerd, we are at ground zero for environmental collapse. We will experience its ill effects first, and most severely. Those effects are already impossible to ignore: strong, prolonged storms flood the campus, closing roads and parking lots while confused catfish work their way onto lawns nearby. Invasive species, some benign, some catastrophically out of tune with the local environment, appear on what seems like a weekly basis. Sinkholes caused by the ongoing depletion of the Floridian aquifer periodically swallow homes in the northern part of the county. Huge dredges are a frequent sight offshore, pumping millions of cubic yards of sand onto the Gulf beaches depleted by ever-stronger and more frequent weather events. Mere weeks prior to the writing of an early draft of this chapter, in August 2015, the City of St. Petersburg's wastewater facilities, inundated by weeks of record-breaking rain, dumped over 15 million gallons of untreated sewage into Clam Bayou, which adjoins the college.[19] The influx of sewage shut down Eckerd's waterfront and sent eighteen-inch-deep streams of effluent flowing in front of the administration building. This is what the early stages of environmental disaster look like, and we have a front-row seat.

Yet, as the old adage about crisis and opportunity goes, I believe that Eckerd's proximity to environmental threats places its faculty in an excellent position to demonstrate those threats to our students and help conceptualize ways in which they might be alleviated. When coupled with a strong sense of place and regional attachment, the visibility of that place's endangerment can inspire powerful action. Our position at ground zero may even offer us the possibility for very real leadership in the field of environmental reform, both in the practical terms of campus preservation and sustainability and the formation of policies, pedagogies, and ideologies that might serve as a model for other campuses. It could be argued that this benefit does not outweigh its associated threats—that it would be utterly myopic to claim that being located next to a Superfund site or nuclear test range is a blessing in disguise owing to the potential for more effectively gaining knowledge about these menaces. However, discovering the possibility for parlaying looming environmental threats into curricular opportunities is an essential skill for educators who teach sustainability to cultivate, since, as we are all aware, the situation is apt to get worse before it gets better.

Consider again the commencement address with which I opened this essay. What would convince the students (and faculty) in attendance that McKibbens's warning that day was more than what its context demanded—a ritualistic set of phrases designed to praise the accomplishments of the graduates within a very limited time frame? What could be said—over the course of that speech or over the short four years that an undergraduate spends under our tutelage— that might turn despair and paralysis into hope and action? Our answer is time-sensitive: the sun is climbing, there's a hot day ahead, and the AC is already blasting.

NOTES

1. Bill McKibben, "Commencement Address," Eckerd College, St. Petersburg, FL, May 19, 2013.
2. David W. Orr, *Down to the Wire: Confronting Climate Collapse* (New York: Oxford University Press, 2009), 177. For an example of practical action universities might take, see the Coastal Fund project of the University of California, Santa Barbara, which makes resources available for students to undertake service programs as diverse as moviemaking, habitat restoration, and whale watching, available online at https://coastalfund.as.ucsb.edu/about-us/.
3. David Sobel, *Place-Based Education: Connecting Classrooms and Communities* (Great Barrington, MA: Orion Society, 2004), ii; C. A. Bowers, *Educating for an Ecologically Sustainable Culture: Rethinking Moral Education, Creativity, Intelligence, and Other Modern Orthodoxies* (Albany: State University of New York Press, 1995), 1; United Nations Educational, Scientific, and Cultural Organization, *Guidelines and Recommendations for Reorienting Teacher Education to Address Sustainability*, Technical Paper No. 2, ed. Charles Hopkins and Rosalyn McKeown (Paris: Section for Education for Sustainable Development, Division for the Promotion of Quality Education, October 2005), 59.
4. Mark S. Cladis, "The Culture of Sustainability," in *Teaching Sustainability: Perspectives from the Humanities and Social Sciences*, ed. Wendy Peterson Boring and William Forbes (Nacogdoches, TX: Stephen F. Austin State University Press, 2013), 43; Suzanne Savanick Hansen, "Reflections from a Classroom That Uses the Campus as a Sustainability Teaching Tool," in Boring and Forbes, *Teaching Sustainability: Perspectives*, 262.
5. Governor Scott has been a longtime denier of climate change; during his gubernatorial campaign in 2011, he stated that he had "not been convinced that there's any man-made climate change." See John Van Beekum, "In Florida, Officials Ban Term 'Climate Change,'" *Miami Herald*, March 8, 2015; Editorial, "The Rick Scott Record: An Environmental Disaster," *Tampa Bay Times*, September 5, 2014; Craig Pittman, "Once a Major Issue in Florida, Climate Change Concerns Few in Tallahassee," *Tampa Bay Times*, May 15, 2011.
6. Eudora Welty, "Place in Fiction," in *The Eye of the Story: Selected Essays and Reviews* (New York: Random House, 1978), 128, 118, 123. It is worth noting that Welty discusses only positive feelings at length, "the warm hard earth underfoot, the light and lift of air, the stir and play of mood, the softening bath of atmosphere that give the likeness-to-life that life needs" (128). For another basic perspective on place, see H. L. Weatherby and George Core, *Place in American Fiction: Excursions and Explorations* (Columbia: University of Missouri Press, 2004), who are insistent that place be more than a simple backdrop to

the action of the story: "Place is not mere setting in the sense of a static background but an essential constituent in what Andrew Lytle deems enveloping action" (11).

7. Anthony Giddens, *The Consequences of Modernity* (Redwood City, CA: Stanford University Press, 1991), 19; Barbara Ladd, "Faulkner, Glissant, and Creole Politics," in *Faulkner in Cultural Context: Faulkner and Yoknapatawpha*, ed. Donald Kartiganer and Ann J. Abadie (Jackson: University Press of Mississippi, 1997), 201; Bruce Robbins, "Actually Existing Cosmopolitanism," in *Cosmopolitics: Thinking and Feeling beyond the Nation*, ed. Pheng Cheah and Bruce Robbins (Minneapolis: University of Minnesota Press, 1998), 3; Edward S. Casey, *The Fate of Place: A Philosophical History* (Berkeley: University of California Press, 1997), 333. In a nearly conspiratorial tone, Robert Kern in "Ecocriticism: What Is It Good For?," in *The ISLE Reader: Ecocriticism, 1993–2003*, ed. Michael P. Branch and Scott Slovic (Athens: University of Georgia Press, 2003), claims that faculty members feel pressured by their profession to downplay the importance of place and thus "frequently set aside the literal reality of place or environment that they encounter in their reading—writing off that reality as a textual construction or effect and focusing instead on what they take it to encode" (258).
8. Tim Cresswell, *Place: A Short Introduction* (Malden, MA: Blackwell, 2004), 49; Arturo Escobar, "Place, Nature, and Culture in Discourses of Globalization," in *Localizing Knowledge in a Globalizing World: Recasting the Area Studies Debate*, ed. Ali Mirsepassi, Amrita Basu, and Frederick Weaver (Syracuse, NY: Syracuse University Press, 2003), 37; Yi-Fu Tuan, *Space and Place: The Perspective of Experience* (Minneapolis: University of Minnesota Press, 1977), 54.
9. Sharon M. Meagher, "Building a Pedagogical Toolbox: The Nuts and Bolts of Infusing Sustainability into Humanities and Social Sciences Courses," in Boring and Forbes, *Teaching Sustainability: Perspectives*, 83; Frederick O. Waage, *Teaching Environmental Literature: Materials, Methods, Resources* (New York: Modern Language Association of America, 1985), 114; David A. Gruenewald, "Place-Based Education: Grounding Culturally Responsive Teaching in Geographical Diversity," in *Place-Based Education in the Global Age: Local Diversity*, ed. David A. Gruenewald and Gregory A. Smith (New York: Routledge, 2007), 143.
10. Gruenewald and Smith, *Place-Based Education*, xx; Steve Chase, "Changing the Nature of Environmental Studies: Teaching Environmental Justice to 'Mainstream' Students," in *The Environmental Justice Reader: Politics, Poetics, and Pedagogy*, ed. Joni Adamson, Mei Mei Evans, and Rachel Stein (Tucson: University of Arizona Press, 2002), 350.
11. The Eckerd campus itself, in fact, is partially a product of this brand of opportunism; roughly half of the land that the college now occupies was dredged out of the

surrounding Boca Ciega Bay at the time of its foundation. The recently constructed $7 million James Center for Molecular and Life Sciences bears a prominent dividing line on the floor of its lobby that demarcates the original shoreline prior to the landfill. It is difficult to walk this line without a creeping sense of foreboding.

In personal terms, I have spent the vast majority of my life occupying landscapes that share with Florida a legacy of despoliation and extraction, landscapes that are, in a word, distant from what is popularly conceived of as "normal" or "civilized." I grew up on a homestead clinging to a wall of the Matanuska-Susitna Valley in south central Alaska, a veritable wilderness when I arrived, now a rapidly suburbanizing and increasingly crowded area (albeit one still cratered with huskies chained to oil drums and half-subterranean galvanized cabins displaying indeterminate game on meat hooks). The Arctic, writes Barry Lopez in *Arctic Dreams: Imagination and Desire in a Northern Landscape* (New York: Charles Scribner's Sons, 1986), is a place where "people's desires and aspirations [are] as much of a part of the land as the wind, solitary animals, and the bright fields of stone and tundra" (xxii).

The same could be said of Florida—its populated history has been shaped to an enormous extent by what its residents have hoped to find in its dim bayous: rewards as exotic as the Fountain of Youth or as mundane as a sunny place in which to play golf and grow old. In both cases, the land suffers the burden of imagination and the struggle to bring those dreams to fruition.

12. Michael K. Stone and Zenobia Barlow, *Ecological Literacy: Educating Our Children for a Sustainable World* (San Francisco: Sierra Club Books, 2005), 93; Gruenewald and Smith, *Place-Based Education*, xvi; Mark DiMaggio, "Educating for Sustainability in American High Schools," in *Education for a Sustainable Future: A Paradigm of Hope for the 21st Century*, ed. Keith A. Wheeler and Anne Perraca Bijur (New York: Springer, 2000), 76.
13. Bowers, *Educating for an Ecologically Sustainable Culture*, 11, 27–28; John Blewitt, "Sustainability and Lifelong Learning," in *The Sustainability Curriculum: The Challenge for Higher Education*, ed. John Blewitt and Cedric Cullingford (New York: Earthscan, 2004), 28–29.
14. McKibben, "Commencement Address."
15. Lynn Mortensen, "Global Change Education," in Wheeler and Bijur, *Education for a Sustainable Future*, 25; Orr, *Down to the Wire*, 5; Danielle Lake, "Sustainability as a Core Issue in Diversity and Critical Thinking Education," in *Teaching Sustainability/Teaching Sustainably*, ed. Kristen Allen Bartels and Kelly A. Parker (Sterling, VA: Stylus Publishing, 2012), 34.
16. Orr, *Down to the Wire*, xiii, 182; Robert Figueroa, "Teaching for Transformation: Lessons

from Environmental Justice," in Adamson, Evans, and Stein, *Environmental Justice Reader*, 325–326.

17. Jon Jensen, "Learning Outcomes for Sustainability in the Humanities," in Boring and Forbes, *Teaching Sustainability: Perspectives*, 35; Boring and Forbes, *Teaching Sustainability: Perspectives*, x.
18. Catherine O'Brien, "Sustainable Happiness and Education: Educating Teachers and Students in the 21st Century," in Bartels and Parker, *Teaching Sustainability/Teaching Sustainably*, 43; Ryan Anderson, review of *This Changes Everything: Capitalism vs. the Climate*, by Naomi Klein, *Anthropology News* 56, no. 7 (July 2015), n.p.

In "Teaching Sustainable Science," in *Teaching as Activism: Equity Meets Environmentalism*, ed. Peggy Tripp and Linda Muzzin (Montreal: McGill-Queen's University Press, 2005), Peggy Tripp affirms feelings of reluctance to incorporate emotional considerations of environmental change, claiming that she was motivated to address the issue of the emotional content of environmental issues only "when comments about hopelessness escalated to the point of obviating student engagement in course material" (75).

19. Zachary Sampson, "Sewage Pumped into Clam Bayou Places St. Petersburg and Eckerd College at Odds, Again, over Wastewater," *Tampa Bay Times*, August 6, 2015. Perhaps the most notorious of invasive species wreaking havoc in Florida is the Burmese python, whose impact on the Everglades ecosystem has been severe; yet even in the Tampa Bay area, it is impossible to traverse even carefully preserved natural areas without encountering a cornucopia of invaders: the Cuban brown tree lizard, which has forced out the native green anole in many regions; the Brazilian pepper, which crowds out mangroves and encourages shore soil degradation; and the ubiquitous air potato, which robs native species of sunlight and growing space.

BIBLIOGRAPHY

Adamson, Joni, Mei Evans, and Rachel Stein, eds. *The Environmental Justice Reader: Politics, Poetics, and Pedagogy*. Tucson: University of Arizona Press, 2002.

Bartels, Kirsten Allen, and Kelly A. Parker, eds. *Teaching Sustainability/Teaching Sustainably*. Sterling, VA: Stylus Publishing, 2012.

Blewitt, John. "Sustainability and Lifelong Learning." In *The Sustainability Curriculum: The Challenge for Higher Education*, edited by John Blewitt and Cedric Cullingford, 24–42. New York: Earthscan, 2004.

Boring, Wendy, and William Forbes, eds. *Teaching Sustainability: Perspectives from the Humanities and Social Sciences*. Nacogdoches, TX: Stephen F. Austin State University

Press, 2013.

Bowers, C. A. *Educating for an Ecologically Sustainable Culture: Rethinking Moral Education, Creativity, Intelligence, and Other Modern Orthodoxies*. Albany: State University of New York Press, 1995.

Casey, Edward S. *The Fate of Place: A Philosophical History*. Berkeley: University of California Press, 1997.

Chase, Steve. "Changing the Nature of Environmental Studies: Teaching Environmental Justice to 'Mainstream' Students." In Adamson, Evans, and Stein, *The Environmental Justice Reader*, 350–367.

Cladis, Mark S. "The Culture of Sustainability." In Boring and Forbes, *Teaching Sustainability: Perspectives*, 41–60.

Cresswell, Tim. *Place: A Short Introduction*. Malden, MA: Blackwell, 2004.

DiMaggio, Mark. "Educating for Sustainability in American High Schools." In Wheeler and Bijur, *Education for a Sustainable Future*, 73–81.

Escobar, Arturo. "Place, Nature, and Culture in Discourses of Globalization." In *Localizing Knowledge in a Globalizing World: Recasting the Area Studies Debate*, edited by Ali Mirsepassi, Amrita Basu, and Frederick Weaver, 37–59. Syracuse, NY: Syracuse University Press, 2003.

Figueroa, Robert. "Teaching for Transformation: Lessons from Environmental Justice." In Adamson, Evans, and Stein, *The Environmental Justice Reader*, 311–330.

Giddens, Anthony. *The Consequences of Modernity*. Redwood City, CA: Stanford University Press, 1991.

Gruenewald, David A. "Place-Based Education: Grounding Culturally Responsive Teaching in Geographical Diversity." In Gruenewald and Smith, *Place-Based Education in the Global Age*, 137–154.

Gruenewald, David A., and Gregory A. Smith, eds. *Place-Based Education in the Global Age: Local Diversity*. New York: Routledge, 2007.

Hansen, Suzanne Savanick. "Reflections from a Classroom That Uses the Campus as a Sustainability Teaching Tool." In Boring and Forbes, *Teaching Sustainability: Perspectives*, 254–264.

Jensen, Jon. "Learning Outcomes for Sustainability in the Humanities." In Boring and Forbes, *Teaching Sustainability: Perspectives*, 23–49.

Kern, Robert. "Ecocriticism: What Is It Good For?" In *The ISLE Reader: Ecocriticism, 1993–2003*, edited by Michael P. Branch and Scott Slovic, 258–281. Athens: University of Georgia Press, 2003.

Ladd, Barbara. "Faulkner, Glissant, and Creole Poetics." In *Faulkner in Cultural Context:*

Faulkner and Yoknapatawpha, edited by Donald Kartiganer and Ann J. Abadie, 31–50. Jackson: University Press of Mississippi, 1997.

Lake, Danielle. "Sustainability as a Core Issue in Diversity and Critical Thinking Education." In Bartels and Parker, *Teaching Sustainability/Teaching Sustainably*, 31–40.

Lopez, Barry. *Arctic Dreams: Imagination and Desire in a Northern Landscape*. New York: Charles Scribner's Sons, 1986.

McKibben, Bill. "Commencement Address." Eckerd College, St. Petersburg, FL, May 19, 2013.

Meagher, Sharon M. "Building a Pedagogical Toolbox: The Nuts and Bolts of Infusing Sustainability into Humanities and Social Sciences Courses." In Boring and Forbes, *Teaching Sustainability: Perspectives*, 76–88.

Mortensen, Lynn. "Global Change Education." In Wheeler and Bijur, *Education for a Sustainable Future*, 15–34.

O'Brien, Catherine. "Sustainable Happiness and Education: Educating Teachers and Students in the 21st Century." In Bartels and Parker, *Teaching Sustainability/Teaching Sustainably*, 41–52.

Orr, David W. *Down to the Wire: Confronting Climate Collapse*. New York: Oxford University Press, 2009.

Robbins, Bruce. "Actually Existing Cosmopolitanism." In *Cosmopolitics: Thinking and Feeling beyond the Nation*, edited by Pheng Cheah and Bruce Robbins, 1–19. Minneapolis: University of Minnesota Press, 1998.

Sobel, David. *Place-Based Education: Connecting Classrooms and Communities*. Great Barrington, MA: Orion Society, 2004.

Stone, Michael K., and Zenobia Barlow. *Ecological Literacy: Educating Our Children for a Sustainable World*. San Francisco: Sierra Club Books, 2005.

Tripp, Peggy. "Teaching Sustainable Science." In *Teaching as Activism: Equity Meets Environmentalism*, edited by Peggy Tripp and Linda Muzzin, 65–80. Montreal: McGill-Queen's University Press, 2005.

Tuan, Yi-Fu. *Space and Place: The Perspective of Experience*. Minneapolis: University of Minnesota Press, 1977.

United Nations Educational, Scientific, and Cultural Organization. *Guidelines and Recommendations for Reorienting Teacher Education to Address Sustainability*. Technical Paper No. 2, edited by Charles Hopkins and Rosalyn McKeown. Paris: Section for Education for Sustainable Development, Division for the Promotion of Quality Education, October 2005.

Waage, Frederick O. *Teaching Environmental Literature: Materials, Methods, Resources*. New York: Modern Language Association of America, 1985.

Weatherby, H. L., and George Core, eds. *Place in American Fiction: Excursions and Explorations*. Columbia: University of Missouri Press, 2004.

Welty, Eudora. "Place in Fiction." In *The Eye of the Story: Selected Essays and Reviews*. New York: Random House, 1978, 116–133.

Wheeler, Keith A., and Anne Perraca Bijur, eds. *Education for a Sustainable Future: A Paradigm of Hope for the 21st Century*. New York: Springer, 2000.

Cutting through the Smog: Teaching Mountaintop Removal at a University Powered by Coal

Brianna R. Burke

What we are doing to this land is not only murder. It is suicide.

—Ann Pancake, *Strange as This Weather Has Been*

Stepping onto the campus of Iowa State University, you can't help but notice the impressive ancient trees, the dignified brick and stone buildings, the flawless work of what seems like hundreds of groundskeepers, and, if you care about the environment, the enormous energy plant lurking on the edge of it all, a big smog-belching monster. Iowa State University offers a PhD in wind energy and dominates academic research in biofuels, but running the university depends on coal, the dirtiest energy source on the planet. Our power plant is ten stories tall, and emissions float over our campus twenty-four hours a day; yet, amazingly, it might as well be invisible, because many on campus—students, staff, and faculty alike—are virtually incapable of seeing it.

As soon as I arrived in 2011 to teach environmental humanities, I knew I would have to address the haunting behemoth that makes everything on campus possible. On the first day, as I pulled into the parking lot adjacent to my building, the power plant towered above me, framed perfectly by my windshield. Coal is often considered the least expensive energy source on the planet, but it is only because what the industry calls "externalities"—the ecological and social devastation caused by its extraction, and the myriad pollutants shed by its combustion—are not figured into its kilowatt-per-hour price. Instead, those "externalities" burden the global commons and are expensive indeed: coal is the largest contributor to global greenhouse-gas emissions.[1] If we truly hope to slow climate change, we must stop burning coal and leave the rest of it in the ground. We must start educating for sustainability, especially in the environments where unsustainable practices are woven into the fabric of our everyday lives.

Because our contribution to climate change is visible at Iowa State University and because climate change will dominate the future of every student in our classrooms, I designed a unit plan to connect my students to it directly via their energy consumption on campus. I include this unit both in my literature and environment class and in my transnational environmental justice course (alongside a unit on oil). The unit includes the eco-documentary *The Last Mountain*, written, produced, and directed by Bill Haney (2011); the novel *Strange as This Weather Has Been*, by Ann Pancake (2007); an "energy consumption" journal; the song "Stream of Conscience," by 2/3 Goat (2010); and a field trip to the power plant itself. Like many in the environmental humanities, my pedagogy aims to render the invisible impacts of our everyday choices visible for students who live in a system they cannot see and that is not of their design, and my unit on coal has four primary objectives: it asks students to examine the multiple ways their lifestyles depend on energy consumption, to learn about the social and environmental impacts of coal, to link their consumption of energy directly to coal via a trip to the power plant, and, finally, to discuss personal accountability and activism, a theme that runs throughout my courses.

Leveling Appalachia

The Last Mountain is the perfect text to open our discussion of coal and its environmental impacts, for when it comes to mountaintop removal, I have found that the old cliché "seeing is believing" is true. You can tell people that coal companies are literally blowing up mountains, that their modus operandi is to plow down some of the oldest and most biodiverse forest in the world, push it over the edge of the mountain, detonate 2,500 tons of explosives "equivalent to the power of a Hiroshima bomb every week," sift through the rubble for coal seams, and then dump the rest off the edges of what has become a barren wasteland into the valleys and streams below, where communities have lived for more than seven generations.[2] But until people actually *see* an entire mountain heave with TNT and disintegrate into a pile of rock and dust, they won't believe it. Words cannot encompass the sheer scale and madness of the destruction. The documentary initiates multiple conversations that echo throughout the course, encompassing our resource pleonexia, its effect on environments and peoples in other places, and its impact on us; the corruption of the democratic process by corporate power and financial influence; climate change and the mandate for clean, sustainable energy; grassroots activism at community, state, and national levels; labor laws, personal rights, and corporate "accountability"; the delicate "insider/outsider" balance of environmental activism; and what it means to love place or home so much that those who belong to it will fight desperately to save it. In fewer words, the film begins a larger and longer conversation about environmental justice and how all of our bodies inhabit complex networks of place, politics, and ecological processes—what Stacy Alaimo calls "transcorporeality."

The documentary opens with an aerial view of the mountains while a hawk circles overhead. As the camera pans, a cloud of smog issuing from a coal-burning power plant oozes across the screen, blocking our view of the forest. Bill Raney, president of the West Virginia Coal Association, says in a voiceover: "I don't think people understand where electricity comes from. I think most people feel like it's an entitlement." We then are told that "almost half of the electricity in the United States comes from burning coal" and that the film will revolve around the fight to save Coal River Mountain in West Virginia.[3] I linger over these opening moments, discussing the jarring juxtaposition between deciduous forest and pollution; I then ask my students: Is Bill Raney right? Is energy an entitlement? I want students to see that they live as if energy *is* an entitlement—we are taught to live our lives this way, and the environmental price of our lifestyles is purposely obscured. In this way, students' lives are bound into a system they did not create but nonetheless are culpable in perpetuating. Most of us are. The realization that they are complicit in destroying the environments and the lives of others without knowing it is devastating to many students; they can get stuck here, frozen in place. I understand their frustration. I feel it, too. But first we must learn to *see* the systems that co-opt our consent to conduct practices we may find abhorrent (or even criminal)—only then can we understand our place within them, begin to think of ways to work around them, and create strategies for enacting large-scale change.

The Last Mountain gives many examples of people doing just that—fighting to change a system, both from inside Appalachian communities and from outside. The documentary introduces us to Maria Gunnoe, resident of Coal River Valley and recent winner of the Goldman Environmental Prize; to Bo Webb, a member of the Coal River Mountain Watch; to Jennifer Massey, who explains how water contamination has affected her small community; and to Ed Wiley and his granddaughter as they fight to relocate Marsh Fork Elementary School, imperiled by an enormous slurry impoundment of toxin-laden sludge. These people are our local, on-the-ground grassroots community activists; they are the voices and faces of the fight against mountaintop removal, making it personal and intimate. Director Haney pairs these activists with "outsiders," most notably Bobby Kennedy Jr. and members of Climate Ground Zero, to show that the fight is not just a local problem, but truly national in scope.

My discussion of the role "outsiders" can play in working with a community for change developed through successive years of teaching this unit, when I found students asked how they could get involved. To talk about the role of "outsiders," I begin by discussing what David Aaron Smith, an activist for Climate Ground Zero, says in the beginning of the film: "I feel like there is a huge crime going on, and I have a right as a citizen to get in the way of a crime taking place." I turn to my students: Do you agree? Do you ever feel that way? Why or why not? As a class we contrast Smith's statement with Chad Stevens's critique of Kennedy's inclusion in the film. In a review of the documentary in the *Journal of Appalachian Studies*, Stevens argues that while Kennedy is a "great character" whose "celebrity" will draw attention to both the film and mountaintop removal, "the focus on Kennedy also places the people who are sick and struggling in the shadow of this one man, an outsider who has, presumably, come to the rescue of the locals."[4] Is this critique fair, I ask students?

What Stevens really addresses with his criticism of Kennedy's inclusion in the film is the difference between fighting *alongside* a community versus purporting to speak *for* them, a distinction I emphasize to my students. The people who live in the shadow of mountaintop removal don't need a savior, I tell my students; they need *allies*. There is a huge difference between the two. We need to start educating students on the ways they can enact their education in the world and on what it means to be an ally. As allies, "outsiders" can garner national attention to a problem often ignored or hidden, which is what I also see myself doing. Like many of my students, I do not want to be complicit in a system that exploits others while silently co-opting my consent to do so. Rendering the otherwise invisible impacts of our energy consumption visible for students is one of the ways I address my own personal complicity, but there are other ways to address it as well. As the unit lesson unfolds, we discuss the personal changes they can make by being mindful of their energy consumption. We also discuss how they can get involved by joining protests, because "outsider" bodies can make certain kinds of violence against "insider" activists unacceptable. If we aren't willing to put our bodies into the fight, we can always agitate our politicians for change or lend our support to local people working on the ground in any way possible. *The Last Mountain* will suggest avenues for action, too, while showing that the balance between "insider" and "outsider" environmental and social activists fighting mountaintop removal is delicate. Both groups of activists are often scorned by the very people they want to protect—those for whom mining is a way of life and, in most of rural West Virginia and the rest of Appalachia, the only work available.

Unfortunately, the film does not give those who work for coal a real voice, an omission that is problematic, as it replicates their loss of voice in the industry as a whole. Students can't imagine the lives of these men and women or the pressures they live under, which is why *Strange as This Weather Has Been* is an important corollary to the film. Instead, in *The Last Mountain*, coal workers are portrayed as ignorant, manipulated, "white trash" hicks. They appear briefly in a few scenes as they protest anti-mountaintop-removal activists, shouting, "Go home" and "Our coal, our jobs." We even glimpse a Massey Coal–sponsored rally where workers are encouraged to engage in a call-and-response chant:

> Whose coal? *Our coal!*
> Whose mountains? *Our mountains!*
> Whose jobs? *Our jobs!*
> Whose freedom? *Our freedom!*
> Whose America? *Our America!*

Don Blakenship, then the chief executive officer of Massey Coal, subsequently takes the stage and tells workers that "Environmental extremists are all endangering American labor. In fact, they are making American labor the real endangered species!"[5] While it is true that viewers are given a small window into how coal workers are manipulated and intentionally made to fear environmental activists (in the chant above, "coal" equals "jobs" equals "freedom"), what is missing is any sense of compassion for these men or their actual stories. Students don't understand that these men have families to support and labor under generations of pressure that have taught them that their very identities as men are linked to working with coal; furthermore, they have little, if any, power over their labor conditions. Massey Coal and other coal conglomerates have demonstrated time and again that their workers are fully replaceable. Students often will ask questions like "why don't they just move?" without comprehending the depth of poverty in the region (to where? with what money?), which directly corresponds to wealth extracted in coal, or that educational attainment is also lowest in areas with the most environmental destruction.[6] In addition, because most of our students have come of age in a culture that does not value the environment or land in any other but economic terms, they forget to weigh what it means to belong to—and love—place. Most of my students are from Iowa or the Midwest, so of course they also don't know the long history of labor, gender, and environmental politics in Appalachia, a gap *Strange as This Weather Has Been* subsequently helps to fill.

The Last Mountain finishes by including all viewers in the fight against mountaintop removal through a long discussion of climate change. Coal consumption doesn't affect the people and the environments in Appalachia alone. It affects everyone. Because "coal is the largest source of greenhouse gases worldwide," the documentary declares that "the epicenter of the climate change battle is Appalachian coal." It then shows viewers how much the coal lobby spent on political campaigns—$86 million—and argues that sustainable economic development in the region could instead reside in wind energy.[7] I rest here with my students for a moment to discuss both corporate influence in our democratic system and wind energy as a viable solution to our energy needs. In 2012, the coal lobby spent more than $153 million on political campaigns, donating 89 percent of their money to Republicans. Alpha Natural Resources, which acquired Massey Energy in 2011, was the largest contributor by far. As for wind energy, Iowans have a great deal of pride in our state's burgeoning industry, which provides more renewable clean energy to its citizens than any other state in the nation.[8] If my students didn't exactly know that investments in sustainable green energy bring economic prosperity, they can *feel* it as more people move into the state to work in the wind energy industry. We will return to these issues again after we visit the energy plant on campus, because even as it is true that Iowa is on the forefront of renewable wind energy development, our infrastructure at ISU still depends on coal, and so perhaps here more than elsewhere, Gunnoe is right when she ends *The Last Mountain* by declaring, "You're connected to coal, whether you realize it or not. Everyone is."

While *The Last Mountain* opens a panoply of issues we will slowly work our way through for weeks, one of the weakest aspects of the film is its failure to discuss love of place, or *topophilia*. Perhaps such a discussion would bleed into sentimentality, but it is a deep, unspoken undertow to the narrative of the documentary. In class, I have students write and share a series of free responses to the question of whether they feel connected to a place. Some do, having grown up on small family farms or in families who value a connection to land, while others feel no connection whatsoever. It is important to discuss these broader cultural ideologies, I think, because more and more people are taught not to cultivate a relationship with place. Indeed, the economy they will enter after graduation demands transiency, for many of my students must relocate to find viable work. But love of place is precisely what leads Marie Gunnoe, Bo Weber, Jennifer Massey, Ed Wiley, and countless others we do not see to fight. This is their home at stake, and "home" is bound to family, futurity, and ecological continuance. Of *The Last Mountain*, Stevens concludes:

> This kind of storytelling is like a stone skipping across the surface of a body of water. It catches our eye, demands our attention and encourages us to question the issue. To continue the dialog, we need stories of the people deeply affected by the gigantic politics and economics of the region . . . and we need empathy for the people.[9]

I agree, and I have seen through successive groups of students each and every year that empathy and compassion can be taught—or at least cultivated—through education. So I continue our conversation through Pancake's novel *Strange as This Weather Has Been*, which explores the tragedy of mountaintop removal through seven primary characters.

The Horizon, Gone

As a multigenerational novel, *Strange as This Weather Has Been* demonstrates how the politics and near-total power of coal in Appalachia is ultimately about futurity. The novel follows the Peace family—Lace and Jimmy Make, the parents, as well as their four children, Bant, Dane, Corey and Tommy—as they face the indeterminacy of a poisoned future that might claim their lives. When the novel begins, they have already experienced a catastrophic flood caused by a collapsed sediment pond, and the novel follows their efforts to find out exactly what is happening above their home. Meanwhile, the coal companies lay off workers, cordon off areas of the mountain, arrest residents for "trespassing," and fail to maintain their other sediment ponds, which remain a haunting threat as the mountain reverberates with TNT and flyrock rains down from above. By the end of the novel, the family is broken. Jimmy Make leaves, taking his remaining sons with him, leaving Lace and Bant behind. The novel is not only about environmental injustice; it also illustrates that the violence the coal companies use to extract resources extends equally to the inhabitants of the area, who are viewed as yet another expendable resource.

Strange as This Weather Has Been beautifully articulates the push and pull of belonging to place, the relationship of inhabitation. Sometimes, against our very wills, a place claims us, declares that we belong to it. Many of us don't learn this lesson until a place is "lost," either through corporate exploitation, as is the case with Appalachia, or through dis/relocation. I try to get students to grapple with the fact that "belonging" to place is neither easy nor simple. The characters we meet as we read struggle with the desire to leave, then experience an attendant achy missing when they do. Pancake has remarked in interviews that "the characters both love their places and hate them, need to leave and need to stay."[10] Students can begin to access the characters' feelings by considering their own feelings toward home, places where they belong, yet need or want to leave in order to grow into adulthood. But imagine, I tell them, that instead of wanting to leave home, you are forced out, your family threatened with nowhere to go. Imagine your home disappearing beneath you even as you inhabit it—what Rob Nixon calls "stationary displacement" in *Slow Violence and the Environmentalism of the Poor* (2013). Some of them already know the violence of this kind of displacement and will talk about the loss of their family's small farm to agribusiness. Many of Pancake's characters, including her two primary narrators, Lace and Bant, experience these feelings, and one narrator, Avery, expresses the complex emotional consequences of having left permanently.

It is fitting that Lace and Bant narrate the bulk of the story—they have the strongest connection to their place and so stay behind to fight when Jimmy leaves at the conclusion of the novel. Lace and Bant's situation merely reflects the reality of the fight against mountaintop removal and many other environmental justice fights around the world: they are led by women who are left to hold everything in place. In struggles like these, men are purposely disenfranchised by a system that defines their masculinity in terms of being the primary wage earners; when they are injured or lose their jobs, their very identities as men are threatened. In Appalachia, "shifts in economics . . . cause a kind of emasculation," Pancake notes in an interview.[11]

Although *Strange as This Weather Has Been* is definitely an ecofeminist novel, Pancake has a great deal of sympathy for men who have no alternative options for meaningful work other than coal mining, which from the very beginning of their lives is continually portrayed as a noble and brave profession. In fact, we talk about how the enormity of the loss the people in Appalachia are experiencing is hard for any film or documentary to encompass fully; perhaps it does take 370 pages to begin to voice what it means to lose home, history, community, faith, land, hope, and, often, most of your family in the process. Literature has narrative possibilities and depths that films do not, and one of the most important gaps in the documentary that the novel addresses is the reality of the men who work in the coal industry. For example, like so many men in the novel and throughout Appalachia, Jimmy Make is "injured on the job." He cannot support his family, and this injury wounds more than his body; it ruptures his very identity. In turn, Lace becomes the primary "breadwinner" and then a budding ecofeminist activist.

Added to the narrative complexity of the novel are the viewpoints of the four children, and students identify best with these voices. Bantella (or Bant) is the oldest child, a teenager for most of the story. She is born at a time when the family is very poor and depends on the mountains for food, so she learns an impressive breadth of local ecological knowledge and forms a deep attachment to place. Students identify with Bant because the novel is partially her coming-of-age story. Dane is the next oldest, and he is marked by a nonspecific birth defect (caused perhaps by mercury and pollution); his body is feminized and underdeveloped, and he suffers from a recognizable anxiety disorder. He says, "I am only twelve years old. And I'm going to see the End of the World."[12] Corey is the next sibling in birth order and the stereotypical white American boy, obsessed with machines and the mine-waste flood-trash that litters the valley. He feels no connection to the land and sees the "hollow" as a narrow, limiting place. Last, there is Tommy, so young and innocent that his character's voice is as underdeveloped as he is. When the flood hits, Tommy sits on the kitchen floor and cries for the family's dogs.

All of the children live with fear as their constant companions; they sense it from their parents, from the slow disappearance of species around them, from the very glass- and coal-dust-filled air. It is arguable that since bombs are constantly exploding overhead, all of the characters in *Strange as This Weather Has Been*—and perhaps many people in Appalachia as well—suffer from post-traumatic stress disorder that is not "post," as would anyone living in a war-torn and threatened country. Fear changes all of them. Corey dies, technically in a four-wheeling accident, but as Carmen Rueda-Ramos notes, "it is machinery that kills him in a land that has been slaughtered and transformed into a dangerous vertical spillway wall where children play."[13] Jimmy Make, Dane, and Tommy become refugees. Lace and Bant turn fear into anger, and then anger into radicalization.

Pancake also gives us two more pivotal narrators, Avery and Mogey, whose chapters act as hinges of different kinds. Avery is a representative of the many who leave Appalachia searching for education or jobs—or any semblance of life that doesn't revolve around coal mining—and never return, but who nevertheless leave behind family that ties them to place. Pancake uses Avery's story to recount the Buffalo Creek Disaster of February 26, 1972, when 125 people died, more than one thousand were injured, and more than four thousand were left homeless because a slurry impoundment flooded the hollow with 132 million gallons of a toxic, gelatinous, poisonous substance that one activist in *The Last Mountain* calls "coal toxin concentrate." In turn, if Avery is Pancake's way of encapsulating history and longing, Mogey's voice is the soul of the novel. He is described as having "the gentleness of trees."[14] Injured when a kettle bottom fell on his head, he has debilitating headaches, but also an insight into place that is profound.

As students watch *The Last Mountain* and then read *Strange as This Weather Has Been*, I ask them to keep a running tally of all of the "externalities" caused by mountaintop removal; by the end of the novel, the list is overwhelming. *The Last Mountain* does a good job of isolating the major issues—water pollution, flyrock and dust, cancer, climate change—but *Strange as This Weather Has Been* fills the outlines of the picture in with painful detail: cracked foundations, lost jobs, disappearing food sources, dying and migrating species, spoiled wells, physical injuries, fear campaigns, intimidation, secrecy, the use of eminent domain to prevent citizens from knowing what is above their heads, and slowly dying communities, drained of citizens and cultural capital. Pancake accurately labels this a "cultural genocide."[15] The list, covering the entire board, shows how, bit by bit, the people in Appalachia are losing their land, their homes, their souls, and even their very lives, because they have the misfortune of living in an area with a natural resource.

Personal Accountability

Pancake's novel is a commitment—at 370 pages, it takes about three weeks to discuss thoroughly. The last week as we finish *Strange as This Weather Has Been*, students keep a five-day journal recording their energy consumption. As they keep these journals, we discuss in class the ways they consume energy that they do not realize, such as when they eat or simply by living within the campus grounds, which are magically manicured by an invisible army of gardeners using all kinds of machines. Did they go to the grocery store? Food requires transportation, which uses energy; the lights in the grocery store use energy; getting to the grocery store means using more energy (unless they bicycled there); cooking food requires yet more. Did they listen to an iPod? Energy. Use their computers? That question is almost laughable. Phones? Clothing? Every single day, almost everything we do other than interact with one another or read outside requires energy and so has an environmental cost. We list the activities on the board, separating them by probable energy type: coal (if on campus), some wind, batteries (and lithium has its own problems), natural gas (ditto), and oil. The vast majority of the energy they use comes from the Iowa State energy plant. Energy consumption is so completely intermeshed with how we live our lives that we no longer can imagine how to live without it. We have become extensions of a gluttonous energy-consuming monster. By the time we are ready to visit the power plant, my students are thoroughly riled, ready to denounce coal and demonize anyone who works for it.

Our tour of the power plant changes everything. I hadn't realized until writing this chapter that this experience is probably the least replicable element of my coal unit due to liability issues, but it is worth attempting to tour your local energy plant because this part of the unit is more powerful than I could have imagined. First of all, walking into an energy plant is humbling. You walk into a huge room reverberating with machinery, *millions* of moving parts, and you can't help but be amazed at the ingenuity of humanity. There is a sense, based in the very arrogance that drives our resource consumption, that if we can build this, we can build anything. There is a little flash of hope, too: we figured this out, so maybe we can actually do better.

Thanks to *The Last Mountain* and *Strange as This Weather Has Been*, students at this point have created a binary where good, honest people are on one side and everyone associated with coal is on the other. However, when faced with the men who work at our coal plant, that binary ruptures, creating an opportunity to talk about the gap between the people who run the company and the people who work for it. In talking with the men who staff our coal plant, it is clear that they are proud of their jobs. Each guide talks about what a strong "environmental" record our plant has, and while we might think such a statement is a contradiction in terms, the men who work at Iowa State's energy plant take it seriously. They have conducted independent studies to ascertain if the waste from the plant enters the groundwater—the university produces 2.7 tons of ash every hour—and have written their own environmental protections for disposal of this waste, because before they did, there weren't any.[16]

But easily the most valuable part of the tour is how it shows students that they are complicit in the practices of Big Coal simply because of how and where they live, and that this was not a choice they were allowed to make—their consent was co-opted. On our last trip to the plant, our guide stood in front of the conveyor belt that carries coal to the burners at a terrifying pace and remarked to the students, "When you flip on a light switch in your dorm room, this conveyor belt speeds up and adds more coal to the fire," and the looks on the students' faces—mixes of fear, wonder, and disgust—were amazing.[17] I couldn't have planned it better had I given him a script.

That night, students write a reflection paper on their energy journal and on what they have learned. It is the place where they can vent and give voice to feelings of guilt; almost all of them feel hopeless and ask what they can do. To open up a conversation to answer that question, which we have spent all semester circling, I use the music of 2/3 Goat, a "Metrobilly" band whose album *Stream of Conscience* brings attention to mountaintop removal and the fight for Blair Mountain in West Virginia.[18] I first incorporated 2/3 Goat because I like to use different kinds of texts in my classroom, especially music—it's a medium as natural to our students as breathing. What's more, it turned into a productive conversation with Annalyse McCoy, the lead singer of the band. She was happy to share her lyrics and discuss with students over Skype why she wrote the title track and what she hopes it will do. She encouraged students to sign a petition on Facebook to save Blair Mountain and to spread awareness of mountaintop removal to their peers, which many do at the end of the semester. But the music also initiates a conversation about the power of art to create social change, using your particular talents—whatever they may be—to work for what you believe in. The band 2/3 Goat does this, as do Pancake, Bill Haney, and every artist we encounter over the course of the semester. If students believe in change, all they have to do is look to our syllabus to find they are not alone.

Making a Case for Hope

Over the time I have taught this unit, some things have changed. Unfortunately, Coal River Mountain was destroyed. However, the campaign to move Marsh Fork Elementary School was successful and (with a lot of pressure) Massey Coal helped pay for it. In addition, there is some hope that Massey Coal may yet pay for its transgressions in another precedent-setting and legally binding way. In October 2015, former CEO Don Blankenship stood trial for misconduct, conspiracy to violate mine-safety standards, and conspiracy to impede federal mine-safety officials. Specifically, he was indicted because of the April 5, 2010, explosion at Upper Big Branch mine, which killed twenty-nine men. The mine was cited for safety violations five hundred times in 2009 alone, and the case shed light on years of irresponsible and reckless (dare I say criminal?) behavior by the coal industry. Yet, the jury in Blankenship's case returned a guilty verdict on only one charge, that of "conspiring to violate mine safety and health standards." On the other two charges, he was declared innocent.[19] In other words, for making the decisions that led to the deaths of twenty-nine men, Blankenship faces at most one year in prison and some fines. (At the time of this writing, Blankenship's appeal of the decision was denied and he has begun serving his sentence.)

Some things have changed on our campus as well, although unfortunately none of the changes were caused by altruism; instead and unsurprisingly, they were motivated by economics. Although Iowa State once purchased its coal from Kentucky, for years it has acquired its supply from Illinois. (Ten percent comes from Colorado.) Due to coal extraction in Illinois, "thousands of acres of the best farm lands and diverse Shawnee National Forests have been left in ruins." Then, in 2013, Iowa State switched two old coal-burning boilers to natural gas. (Maybe now I will have to include in this unit the eco-documentary *Gasland*.) The university also installed a small wind turbine, which is more for student study than actually helping to meet our energy needs: a representative from the power plant remarked that it was "about enough to power the president's home for the year." Over the summer of 2015, it also installed an array of solar panels for instruction and student study, and these add a small amount to the campus energy grid. In addition, last year the university purchased 6.8 percent of the campus's energy from the Story County Wind Farm. When students ask what guided these changes, the answer is, of course, cost, and this is important for students to hear. If we want to win the "energy war" and break our dependence on fossil fuels, we need to be able to price renewable energy resources lower. Already we can see this happening as more and more coal companies declare bankruptcy. Despite these shifts, however, Iowa State in 2014 consumed 102,396 tons of coal.[20] We still hear the coal-laden trains rumbling through town in the middle of the night.

Although it takes nearly five weeks in total, I continue to teach this unit because climate change looms, and every person must be connected to the largest single problem ever to confront our species. Our planet is already hotter, and we know that carbon has a time lapse before its full effects are felt; sea-level rise is worse than was projected, storms are stronger and more destructive, and our ability to grow food or preserve water is threatened. We can no longer afford to be complacent. As I write this in my university office, I can feel warm air issuing from the heater behind me, and I know that more coal has just been added to the fire at the energy plant. I can't change the energy infrastructure at Iowa State, but I can work toward changing the infrastructure of society at large by making the environmental impact of their lives visible for students, for I have found that once they know it, this knowledge haunts them. They want something better, to live in a system less destructive. So do I.

I have had students tell me that after my course they cannot blithely turn on every light in their rooms or hear a train bringing coal to the university in the middle of the night without thinking about the social and environmental costs of the way they live their lives. They often tell me they are trying to make different and, when possible, better and more deliberate choices—but in this case, with energy infrastructure, making large-scale changes is hard. Some changes can be individual or community-based (grassroots or "bottom-up"), but others must be structural ("top-down"). This unit, which appears early in my course, functions in part to draw the distinction between the two kinds of changes needed when it comes to creating more sustainable ways to live. I watch as students struggle to incorporate the personal changes demanded of them, even as they work toward leveraging pressure to alter the more structural problems that face us. But their willingness and desire to change always amazes me. Many people have written about the supposed vapidity of the millennials, arguing that they are self-involved and apathetic—but I have to admit, I don't see it. Each semester I meet another group of young people who look at the truth of climate change, and then think through their own personal contributions to the problem, and who leave my class ready to work for systemic change. I invest much of my hope for the future in their belief that they can make a more sustainable, socially and environmentally just world, and that they can enact the changes demanded of us.

NOTES

1. According to a recent study published in the *Annals of the New York Academy of Sciences*, factoring the "externalities" of coal into the kilowatt-per-hour price would "conservatively" double or triple the cost. In addition, the authors assert, "the life cycle effects of coal and the waste stream generated are costing the U.S. public a third to over one-half of a trillion dollars annually. Many of these so-called externalities are, moreover, cumulative." See Paul R. Epstein et al., "Full Cost Accounting for the Life Cycle of Coal," in *Ecological Economics Reviews*, ed. Robert Costanza, Karin Limburg, and Ida Kubiszewski, special issue, *Annals of the New York Academy of Sciences* 1219 (February 2011): 73–98, http://dx.doi.org/10.1111/j.1749-6632.2010.05890.x; Rob Perks, *Appalachian Heartbreak: Time to End Mountaintop Removal* (New York: Natural Resources Defense Council, n.d.), 1.
2. Perks, *Appalachian Heartbreak*, 2.
3. *The Last Mountain*, dir. Bill Haney (Uncommon Productions, 2011). According to the U.S. Energy Information Administration, in 2014 some 39 percent of domestic energy consumption was produced by burning coal. The film was released in 2011 and likely produced before that, so the figure stated in the film is slightly outdated. See U.S. Department of Energy, Energy Information Administration, "Electricity Explained: Electricity in the United States," last modified March 29, 2016, http://www.eia.gov/energyexplained/index.cfm?page=electricity_in_the_united_states.
4. *Last Mountain*, dir. Haney; Chad Stevens, review of *The Last Mountain*, *Journal of Appalachian Studies* 18, no. 1/2 (Spring/Fall 2012): 317.
5. *Last Mountain*, dir. Haney.
6. Epstein et al., "Full Cost Accounting," 84; Perks, *Appalachian Heartbreak*, 7.
7. *Last Mountain*, dir. Haney.
8. SourceWatch, "Coal Money in Politics," last modified September 20, 2012, http://www.sourcewatch.org/index.php/Coal_money_in_politics; American Wind Energy Association, *Iowa Wind Energy* (Washington, DC: American Wind Energy Association, n.d.), http://awea.files.cms-plus.com/FileDownloads/pdfs/Iowa.pdf. The amount spent by the coal industry cited by *The Last Mountain* ($86 million) is outdated, and it is unclear where this figure comes from.
9. Stevens, review of *The Last Mountain*, 319.
10. Nicholas Arnold and Michael Baccam, "A Conversation with Ann Pancake," *Willow Springs*, April 20, 2007, 13, http://willowspringsmagazine.org/interview/ann-pancake-willow-springs-interview.
11. Ibid., 2.
12. Ann Pancake, *Strange as This Weather Has Been* (Berkeley, CA: Counterpoint, 2007), 112.

13. Carmen Rueda-Ramos, "Polluted Land, Polluted Bodies: Mountaintop Removal in Ann Pancake's *Strange as This Weather Has Been*," in *The Health of the Nation: European Views of the United States*, vol. 6 of *European Views of the United States*, ed. Meldan Tanrisal and Tanfer Emin Tunç (Heidelberg, Germany: Universitätsverlag Winter, 2014), 228.
14. Kai T. Erikson, *Everything in Its Path: Destruction of Community in the Buffalo Creek Flood* (New York: Simon & Schuster, 1978); *The Last Mountain*, dir. Haney; Pancake, *Strange as This Weather Has Been*, 36.
15. Shannon Elizabeth Bell, "'There Ain't No Bond in Town Like There Used to Be': The Destruction of Social Capital in the West Virginia Coalfields," *Sociological Forum* 24, no. 3 (September 2009): 631–657; Arnold and Baccam, "Conversation with Ann Pancake," 8.
16. Jeffrey Witt, interview by author, Ames, IA, March 4, 2012.
17. Ibid.
18. 2/3 Goat, "Stream of Conscience," by Annalyse McCoy and Ryan Dunn, *Stream of Conscience*, 2/3 Goat, 2014, CD.
19. David Segal, "The People v. The Coal Baron," *New York Times*, June 20, 2015; Ken Ward Jr., "Blankenship Guilty of Conspiring to Violate Mine Safety Rules," *Charleston Gazette*, December 3, 2014. See also *Overburden*, dir. Chad A. Stevens (milesfrommaybe Productions, 2015); it tells the story of the Upper Big Branch mine disaster, documenting the battle of two women—one an activist against mountaintop removal, the other formerly pro-coal—to hold Massey Coal accountable for the industrial mining disaster.
20. Mike McGraw, interview by Brianna Burke, Ames, IA, September 2, 2015; Jeff Biggers, "Heartland Coal Crisis: Illinois Bankrolls Big Coal School Program? Interview with Eco-Justice Leader Lan Richart," *Huffington Post*, June 22, 2011, http://www.huffingtonpost.com/jeff-biggers/coal-marketing_b_882383.html; *Gasland*, dir. Josh Fox (Docurama, 2010); McGraw, interview.

BIBLIOGRAPHY

2/3 Goat. *Stream of Conscience*. 2/3 Goat, 2014, CD.

Arnold, Niccolas, and Michael Baccam. "A Conversation with Ann Pancake." *Willow Springs*, April 20, 2007. Http://willowspringsmagazine.org/interview/ann-pancake-willow-springs-interview.

Bell, Shannon Elizabeth. "'There Ain't No Bond in Town Like There Used to Be': The Destruction of Social Capital in the West Virginia Coalfields." *Sociological Forum* 24, no. 3 (September 2009): 631–657.

Epstein, Paul R., Jonathan J. Buonocore, Kevin Eckerle, Michael Hendryx, Benjamin M. Stout

III, Richard Heinberg, Richard W. Clapp, Beverly May, Nancy L. Reinhart, Melissa M. Ahern, Samir K. Doshi, and Leslie Glustrom. "Full Cost Accounting for the Life Cycle of Coal." In *Ecological Economics Reviews*, edited by Robert Costanza, Karin Limburg, and Ida Kubiszewski. Special issue, *Annals of the New York Academy of Sciences* 1219 (February 2011): 73–98. Http://dx.doi.org/10.1111/j.1749-6632.2010.05890.x.

Erikson, Kai T. *Everything in Its Path: Destruction of Community in the Buffalo Creek Flood.* New York: Simon & Schuster, 1978.

The Last Mountain. Directed by Bill Haney. Los Angeles, California: Uncommon Productions, 2011. DVD.

Pancake, Ann. *Strange as This Weather Has Been.* Berkeley, CA: Counterpoint, 2007.

Rueda-Ramos, Carmen. "Polluted Land, Polluted Bodies: Mountaintop Removal in Ann Pancake's *Strange as This Weather Has Been.*" In *The Health of the Nation: European Views of the United States*, vol. 6 of *European Views of the United States*, edited by Meldan Tanrisal and Tanfer Emin Tunç, 219–230. Heidelberg, Germany: Universitätsverlag Winter, 2014.

SourceWatch. "Coal Money in Politics." Last modified September 20, 2012.

U.S. Department of Energy. Energy Information Administration. "Electricity Explained: Electricity in the United States." Last modified March 29, 2016. Http://www.eia.gov/energyexplained/index.cfm?page=electricity_in_the_united_states.

Teaching about Biodiversity and Extinction in a Thawing Alaska: A Reflection

Jennifer Schell

Ecological change is rarely simple, as nature is interrelated and interdependent.

—Alaska Region Climate Change Response Strategy 2010–2014

Alaska's 663,300 square miles contain a diverse array of ecosystems, including temperate rainforests, rocky coastlines, volcanic archipelagoes, alpine meadows, boreal forests, glacial rivers, tidal mudflats, and arctic tundra. Although the Far North is not known for its biodiversity, numerous plant and animal species populate these habitats. As moose browse on willows and beaver fell birches, black and brown bears gorge on blueberries, raspberries, salmonberries, cloudberries, cranberries, and crowberries. Meanwhile, spawning salmon migrate up the Yukon River, Dall sheep scale the slopes of the Alaska Range, caribou roam across the arctic tundra, and polar bears stalk the ice floes of the Beaufort Sea.

All of these plants and animals are deeply affected by climate change, for, as the authors of the National Park Service's *Alaska Region Climate Change Response Strategy 2010–2014* (2010) correctly note, "the effects of climate change are occurring more quickly and with more severity in Alaska than at lower latitudes." Recent scientific studies—the results of which are often published in Fairbanks and

Anchorage newspapers—demonstrate that sea ice is shrinking and permafrost is thawing. Glaciers, too, are melting rapidly. As a result, wildfires are worsening and coastal erosion is increasing. Forest/tundra boundary lines are advancing north, and large amounts of methane are spewing into the atmosphere. Polar bears are interbreeding with grizzlies, producing hybrids called pizzlies or grolars. Displaced walruses are congregating in massive numbers on the remote beaches of northern Alaska; pikas and tundra swans are in danger of losing their habitat entirely.[1] All this evidence indicates that climate change is rapidly altering life in Alaska, often in unpredictable ways.

After spending eight years in Pittsburgh, Pennsylvania, and two years in Wichita, Kansas, I moved to Alaska in 2009 to take a job in the English department at the University of Alaska Fairbanks (UAF), and since my arrival, I have come to appreciate the diversity of ecosystems the state encompasses, just as I have grown increasingly aware of the threats that climate change poses to them. I have also experienced, firsthand, the unsustainability of Alaska's economic dependence on fossil fuel industries. Although my scholarship has always focused on environmental issues—my dissertation and my first book examine early American whaling literature—it now reflects newfound appreciation, awareness, and experience. During the 2014–2015 school year, I began researching a new book manuscript dedicated to representations of biodiversity and extinction in the print culture of the Circumpolar North, the arctic and subarctic regions of the Far North. Because my teaching is directly informed by my research, I also designed and taught a second-year writing class, "Academic Writing about the Social and Natural Sciences," that addressed these issues.

To develop this course, I adopted a form of critical place-based pedagogy and selected a series of twelve readings, some of which addressed Alaskan concerns and others of which addressed national and global issues. I organized these texts into three units—"Science and History," "Science, Politics, and People," and "Case Study of the Polar Bear"—and then composed five writing assignments specifically designed to encourage students to consider the complex interplay of the local, the national, and the global with respect to issues of biodiversity and extinction.

As I taught the class for the first time, I noticed a peculiar issue emerging from considerations of our readings. Several times, my students and I witnessed academic authors and science writers advance alarmist doomsday prophesies. In "Six Biological Reasons Why the Endangered Species Act Doesn't Work—and What to Do about It" (1991), Daniel Rohlf decries the "current precarious state of the earth's biological resources." Meanwhile, in "The Biological Basis for Human Values of Nature" (1995), Stephen Kellert condemns the "contemporary drift toward massive biological impoverishment and environmental destruction." In *On Thin Ice: The Changing World of the Polar Bear* (2010), Richard Ellis laments that polar bears represent humanity's "precarious relationship with nature," and in *Extinction: The Causes and Consequences of the Disappearance of Species* (1981), Paul and Anne Ehrlich predict the possible "extermination" of the human race. For me, these claims raise important questions about the efficacy of apocalyptic rhetoric within the genres of academic and scientific writing. My students, however, were markedly more dismissive of these doomsday prophecies. Among other things, their casual response suggested to me that perhaps apocalyptic environmentalist discourse has lost its power to provoke emotion and action. Part of the problem might be that within the genre of environmentalist writing, alarmist rhetoric has become trite and clichéd, however true it might be. Another issue might be that some provocative predictions—like those advanced by Paul Ehrlich in *The Population Bomb* (1968)—have proven to be wildly inaccurate.[2] Whatever the case, my students remained entirely unmoved by all of the aforementioned authors' claims about environmental devastation.

Unsustainable Environments: Alaska

When I moved to Alaska, I was unaware of the degree to which my livelihood would depend on the fluctuations of fossil fuel revenues. Within weeks of my arrival, I learned that Alaska does not collect a state income or sales tax; instead, it depends on the oil industry to supply approximately 85 to 90 percent of the funds for its operating budget. Of course, this mode of revenue generation is ultimately unsustainable. According to the Alaska Oil and Gas Association, the amount of oil flowing through the Trans-Alaska Pipeline has declined by 39 percent over the past ten years, because of increased production elsewhere in the United States and the difficulty of oil extraction in the Arctic. This problem has been exacerbated of late by falling oil prices. After a protracted and contentious legislative session, the 2015–2016 fiscal year began with the state facing a budget deficit of $3 billion caused by decreases in oil revenues.[3] With a legislature dominated by Republicans who prefer enacting spending cuts to taxes, this situation has had a dire effect on state services, including those furnished by UAF.

As the university struggles to sustain itself under these straitened circumstances, its faculty and staff have learned to cope with layoffs, furloughs, workload changes, and other budget-cutting measures. The latter include the consolidation of campus offices, the reduction of travel budgets, and the elimination of certain degrees, such as the BA in philosophy, the BA in music, and the BS in sociology. Currently, humanities faculty are scaling back their research and creative activities so that they can teach mandated classes and expand their service obligations. Custodians are working day shifts to save electricity. Athletic teams are traveling to fewer events. And construction is grinding to a halt on a new engineering building, leaving it half finished.[4] All of these budgetary issues affect students, who depend on UAF's faculty, staff, and facilities to make their educational experience a generative one.

Despite the unsustainability of the state's revenue stream, many Alaskans remain committed to developing fossil fuel resources. In April 2013, Republican governor Sean Parnell initiated a series of controversial tax cuts intended to increase production in the oil fields on the North Slope. Many Alaskans protested these measures, not because they opposed oil development, but because they took issue with Parnell's dealmaking methods. These state issues have potential national and global implications. All of Alaska's representatives in Washington, DC—Republicans Lisa Murkowski, Dan Sullivan, and Don Young—staunchly support oil drilling in the Arctic National Wildlife Refuge and the National Petroleum Reserve. They have not worked to address the problems posed by climate change. When interior secretary Sally Jewell traveled to Kotzebue in February 2015 to speak with Alaskan Native leaders about climate change and coastal erosion, she was met by Alaska's congressional delegation, who wanted to talk about increasing the state's oil revenues. Neither time nor the recent budget crisis have dissuaded Alaska's political leaders from their dedication to fossil-fuel extraction industries. Just before President Barack Obama's August 2015 visit to Alaska, Murkowski wrote an op-ed piece for Alaska's newspapers in which she argued that "climate change must not be used as an excuse to deprive Alaskans of our best economic prospects," namely, "oil production."[5]

Teaching at UAF: Critical Place-Based Pedagogy

Alaska's special environmental, political, and demographic features make UAF a challenging and an exciting place at which to work. With 8,793 undergraduate and 1,199 graduate students, the university offers upwards of 240 degrees in 179 disciplines, some of the most popular of which are the biological sciences and engineering. Although many of UAF's scientists study climate change—the university is home to the Alaska Climate Research Center and the International Arctic Research Center—its geological and mining engineers endorse development, declaring on their website, "Alaska is a resource state, and its future is linked intricately to the development of its vast land." On their website, the petroleum engineers indicate that they "promote the development and production of conventional and unconventional hydrocarbon resources."[6] Neither site mentions sustainability or climate change.

Together with their colleagues in the humanities and the social sciences, these faculty teach a wide diversity of students. The flagship of the state system, UAF accepts approximately 80 percent of its applicants; thus, any given class typically contains students with a broad range of intellectual abilities and educational goals. Some of these students seek bachelor's degrees, while others pursue associate's degrees, certificates, licenses, or occupational endorsements. Some live on campus in dormitories, while others commute from Fairbanks or the nearby town of North Pole. Some matriculate immediately from high school, while others enroll while, or after, spending time in the work force or military. Some come from villages, such as Huslia or Metlakatla, while others hail from cities, such as Anchorage or Juneau. Some identify as Alaskan Native (18.5 percent), while others identify as Caucasian (45.9 percent), African American (2.1 percent), Asian/Pacific Islander (2.3 percent).[7]

Over the course of my seven years at UAF, I have embraced this diversity, developing an approach to teaching writing that I characterize as a variant of the critical place-based pedagogy described by David A. Gruenewald in "The Best of Both Worlds: A Critical Pedagogy of Place" (2003). Like Gruenewald, I believe that "place-based pedagogies are needed so that the education of citizens might have some direct bearing on the wellbeing of the social and ecological places people actually inhabit" and that "critical pedagogies are needed to challenge the assumptions, practices, and outcomes taken for granted in dominant culture and in conventional education."[8] Fusing these two approaches, I invite my students to explore the particular social and environmental problems posed by the places in which they live and work. Because no community or ecosystem exists in isolation, we also consider how local environmental problems resonate with national and global issues.

One of the difficulties with adopting a critical pedagogy of place stems from issues of nativity, both my own and that of my students. Even though many Fairbanksans attend UAF, the university also draws students from across the state, the nation, and the world. Each group perceives my motives and me somewhat differently. My Alaskan students tend to regard me as an outsider, a rootless academic who has lived in many different regions of the United States. These students are sometimes suspicious of my familiarity with, and reason for, emphasizing the importance of local places, knowledges, and cultural productions. Although some out-of-state students appreciate Alaska, many of them identify more with other places. As such, they are not always interested or invested in local concerns.

To cope with these challenges, I familiarize myself as best I can with the locality in which I live and teach. I talk to local community members, examine local media, investigate local history, research local cultures, participate in local activities, and explore the surrounding region. Although I have always felt at least a tenuous connection to Alaska—my childhood dreams involved working as a naturalist in Denali National Park—these activities strengthen my bond to Fairbanks and inform my teaching. In my "Biodiversity and Extinction" class, I openly acknowledged my outsider status and invited my Alaskan students to help us discern the subtle intricacies of local places and cultures. I shared the knowledge I gleaned from my various activities, and I asked my fellow outsiders to help us think about the complex connections that disparate places share. This approach helped us all to think and write more perceptively about the environmental issues affecting the place in which we live.

Teaching about Biodiversity and Extinction: Spring 2015

On the first day of the 2015 spring semester, I discovered that demographically speaking, this section of "Academic Writing about the Social and Natural Sciences" was unlike others I had taught at UAF. Typically, my classes contain equal numbers of men and women from an array of backgrounds and degree programs. This semester, nineteen of my twenty-two students were young men, most of whom came to UAF directly from high school. As I soon learned, the class was also disproportionately filled with fifteen engineering and computer science majors. Of the remaining seven students, four were pursuing associate's degrees through UAF's Community and Technical College, and three were seeking bachelor's degrees in accounting, political science, and wildlife biology.

Given the number of engineers in the class, I expected some resistance to the environmental materials I planned to cover (thanks to their likely political affiliations), but I did not anticipate the other demographic challenges I encountered. Try as I might to encourage the three women students to join in class discussions, they preferred not to participate. Even when working in small groups, they were fairly quiet. Some of this reticence may be attributed to culture or personality. During midterm conferences, all three of these young women—two of whom identified as Alaskan Native—told me that as naturally quiet people, they found it difficult to talk in class. While they did not mention gender or racial difference, I suspected that these women might have felt intimidated to speak in a room filled predominantly with white men. To ameliorate this issue, I offered them the option to participate during office hours or through email.

Another diversity-related issue that emerged over the course of the semester was that many of the engineering students—especially the petroleum, geological, and mining engineers—endorsed politically conservative viewpoints about resource development. Because so many students possessed similar opinions, they tended to dominate our classroom discussions, and I constantly worried that we did not leave much room for dissenting opinions. Rather than voice these ideas myself, I tried to elicit them from students through small group activities and writing workshops, but I found that, for the most part, the political tenor of the conversation remained largely the same over the course of the semester.

Mindful of, and uneasy about, the demographics of my classroom, I assigned the first readings—selections from Georges Cuvier's "Memoir on the Species of Elephants, Both Living and Fossil" (1796) and Charles Lyell's *Principles of Geology* (1830–1833)—and distributed the first writing prompt. What I appreciate about eighteenth- and nineteenth-century science writing is its long, descriptive sentences and its unconventional forms by today's standards. Thus, I designed the first assignment to encourage students to practice their close reading abilities. After giving my students some historical background and warning them about textual difficulties, I asked them to select a short passage from each reading and compose a summary and analysis of each one. To encourage students to think about different aspects of their quotations, I posed a series of thirteen questions, addressing style, theme, and content. At approximately 760 words, this prompt was, by design, a dense one. My intent was to force students to spend some time wrestling with the prompt, unpacking the prose of the texts, and considering the questions.

After receiving the first drafts of this assignment, I noticed that many students focused their analysis on the scientific content of their chosen passages. Some of these pieces were both ambitious and insightful. One student researched the impact of Christian beliefs on the historical trajectory of extinction studies, and another compared Lyell's observations about the ramifications of extinction to Aldo Leopold's concept of trophic cascades. Still, I wanted students to focus more on Cuvier's and Lyell's language, so in my feedback and our class discussions I urged them to concentrate more on diction, style, and discourse. Once they turned their attention to these features of the readings, they began to notice Cuvier's predisposition toward rhetorical questions and vivid description. Many of them highlighted the paragraph in which Cuvier argues that

> [all] these facts . . . seem to me to prove the existence of a world previous to ours, destroyed by some kind of catastrophe. But what was this primitive earth? What was this nature that was not subject to man's dominion? And what revolution was able to wipe it out, to the point of leaving no trace of it except for some half-decomposed bones?

With respect to Lyell, they emphasized his predilection for random, uncredited Shakespearean allusions (he offhandedly refers to invasive plants as "the darnel, hemlock, and rank fumitory") and his use of the first-person-plural point of view.[9]

Because I also noticed that, for the most part, my students opted to write five-paragraph, thesis-driven essays, I dedicated class time to discussing the advantages and limitations of this mode of writing. I observed that sometimes, students who adopt this approach produce uninspired, formulaic writing. We talked about the benefits of exploratory writing, and we workshopped the papers that experimented with different organizational frameworks. In this way, I encouraged them to take more chances with their writing—and their ideas—as they worked their way through the drafting process.

With the next writing assignment, I invited students to use their newly acquired close reading abilities to grapple with a cryptid controversy that occurred in Alaska at the turn of the nineteenth century. I assigned Henry Tukeman's "The Killing of the Mammoth" (1899), a short story in which the narrator claims to have traveled to Alaska, slayed the titular pachyderm, and donated the carcass to the Smithsonian. In class, I explained that many naive, enthusiastic readers mistook the story for nonfiction, and they began to besiege the museum with letters about Tukeman's mammoth. To alleviate the confusion, Frederic Lucas—a naturalist and taxidermist at the Smithsonian—surveyed the available paleontological evidence and declared the mammoth to be extinct. His essay "The Truth about the Mammoth" (1900) reveals his findings.[10] In the prompt I composed, I asked students to use textual evidence to explain why Tukeman's story was so believable and why Lucas felt so compelled to respond to it. To encourage them to think more about human attitudes toward extinction events, I also asked them to consider the widespread popularity of cryptozoology.

As I suspected, this assignment proved to be particularly popular, especially with rural students familiar with the Alaskan terrain described in the story. Many of them astutely observed that Tukeman establishes his hunting credentials through his claims about shooting elephants in Africa. They also suggested that he renders fantastic descriptions of the mammoth—"the king of the primeval forests"—more believable by prefacing them with realistic portraits of Alaskan geography. For evidence, several students cited the following passage in which Tukeman describes his ascent up the Porcupine River: "Sometimes twice or thrice a day we would unload, to drag our boat over shallows or around log-jams, and on one occasion we had to portage everything a mile overland to avoid a canon. . . . And then the mosquitoes! I have some experience of them, but I have never seen them so bad as they were on the upper reaches of the river during the month of July."[11] In their analysis, they drew on their own expertise, noting that Tukeman's description of interior Alaskan rivers and mosquitoes is fairly accurate.

Of note, several students found themselves distracted by my questions about cryptozoology, writing far more about famous cryptids, such as the Loch Ness Monster and Bigfoot, than they did about Alaskan mammoths. If I were to use this prompt again, I would drop these questions and open up a new line of inquiry, one that encourages students to think about the historical trajectory of popular perceptions of Alaska as a vast wilderness filled with an inexhaustible abundance of charismatic megafauna. I think this approach would be more in keeping with the theme of the course, and it would cater to the interests of Alaskan students, many of whom are intrigued by the various ways in which "reality" television shows, such as *The Deadliest Catch* (2005 to the present), *Ice Road Truckers* (2007 to the present), and *Life Below Zero* (2013 to the present), misrepresent Alaskan life.

After concluding the historically oriented unit of the course, I turned my students' attention to more explicitly political matters. To encourage them to think strategically about modes of argumentation and forms of rhetoric—both their own and those employed by other writers—I asked them to compose a more traditional, thesis-driven essay. As an example of an emotionally driven argument, I assigned the first chapter of the Ehrlichs' *Extinction*, and as an example of a logically driven argument, I assigned Rohlf's "Six Biological Reasons Why the Endangered Species Act Doesn't Work—and What to Do about It." Then, I asked my students to evaluate the efficacy of the arguments advanced by each text.

Significantly, in our discussion of these readings, I noticed what I thought was a decided preference for Rohlf's modes of argumentation. Students complimented his logical, rational tone and his use of supporting evidence. They admitted that Rohlf's concluding statement about the "current precarious state of the earth's biological resources" was emotionally disjunctive, but they seemed willing to forgive this momentary lapse because the remainder of the essay was well argued.[12] When I read through the papers, I learned that several students disagreed with their more vocal classmates. Some of them professed to prefer the Ehrlichs' writing—despite its alarmist tone, doomsday prophecies, and logical fallacies—because it employed vivid anecdotes and sensational imagery. The Ehrlichs were, in short, more fun to read. At the time, I wished that these students had voiced their opinions in class, but I recognize now that I did not necessarily make room for them to do so. Perhaps because I tend to eschew emotional arguments and doomsday prophesies in my own writing, I let the dominant voices in the class—many of them logically minded engineering students—skew the discussion in favor of one author at the expense of the other.

At the midpoint of the semester, I introduced my students to the concept of biophilia, asking them to complete an in-class essay on Kellert's "The Biological Basis for Human Values of Nature." In this prompt, I described the term as controversial because of its association with the field of sociobiology, a discipline whose founding principle is that human behaviors possess genetic origins. Then, I invited my students to explore the problems with the term through their writing. After they completed this portion of the assignment, I asked them to take their essays home and revise them, such that they took into account some of the controversial aspects of Richard Nelson's "Searching for the Lost Arrow: Physical and Spiritual Ecology in the Hunter's World" (1995). One of my goals for this assignment was to help students improve their drafting and revision abilities. In my experience, students who have trouble getting started on assignments often benefit from timed writing, and students who struggle with revision often do more with their papers when they are forced to shift the focus of their topic or consider another text. Another goal was to get them thinking about place-based issues of diversity, especially with respect to Nelson's discussion of Alaskan Native belief systems.[13]

Although this assignment helped students practice drafting and revision, I think that I confused them by asking them to discuss complex concepts not mentioned in the reading or foregrounded in class discussion. During the in-class essay, several students approached me to ask if I could clarify my definition of sociobiology. Others just avoided the term altogether. Another issue stemmed from the fact that I asked them to produce drafts of two essentially different papers for just one assignment. In the future, I might streamline the two prompts into one to eliminate confusion and wasted effort. I also will furnish students with a more legible definition of sociobiology, one that includes a few concrete examples for them to consider.

During the biophilia portions of the course, my students and I really became more cognizant of the apocalyptic trend in our readings. We noticed that, like Rohlf and the Ehrlichs, Kellert employs alarmist language to call for "a fundamental shift in global consciousness—one capable of countering the contemporary drift toward massive biological impoverishment and environmental destruction." And we observed that Nelson attributes Western culture's "imbalance with the environment" and "loss of affinity with life" to its "single-minded pursuit of knowledge and a diminished regard for wisdom."[14] Interestingly, these statements seemed to bother me far more than they did my students. I was insulted by what I perceived to be hyperbolic fearmongering, largely because I believe environmentalists who want to be taken seriously, especially in moderate and conservative circles, should avoid employing unnecessarily alarmist rhetoric. My students, however, were unperturbed.

Since I was so surprised by their response—or lack thereof—to this accumulation of apocalyptic rhetoric, I failed to explore the issue further with them, and so I can only speculate about their thought processes. It is possible that at some point in time they became inured to this kind of language. After all, it is a cliché of environmentalist writing. As Alaskan residents, they have witnessed countless battles between environmentalists and developers over the state's natural resources. Many of these conflicts—including those over the proposed Pebble Mine and Susitna Dam—have yet to be resolved. It is also possible that they have learned not to take doomsday prophecies seriously, because so many of them have gone unfulfilled. Compared to many other parts of the United States, Alaska is visibly filled with wildlife, especially along its southeastern coastline. Moreover, some populations of formerly endangered Alaskan species, including bald eagles and Aleutian Canada geese, have recovered. Likewise, the Trans-Alaska Pipeline has yet to cause the catastrophic environmental devastation that some environmentalists predicted it would.[15] Whatever the reason, I did learn that, at least for this group of students, doomsday discourse no longer has quite the emotional and motivational power that it did in the 1960s, when Rachel Carson published *Silent Spring*.

In the third unit of the course, students composed thesis-driven histories of extinct, endangered, or recovered organisms. Although they selected their own topics, I required them to employ specific kinds of source materials, including at least one published before the twentieth century and three published after the twentieth century. I stressed that all of these sources should be academic and reliable, and I showed them how to conduct research in online databases and the UAF library system. To model the various analytical approaches students might adopt toward their subject matter in their research papers, I engaged them in a chronologically organized case study of polar bear literature. We started with excerpts from Samuel Hearne's *A Journey from Prince of Wales's Fort in Hudson's Bay to the Northern Ocean* (1795) and Frederick Schwatka's *Nimrod in the North* (1885), both of which represent animals as valuable commodities and dangerous adversaries for those hunters daring enough to pursue them. We then turned our attention to chapter 7 of Ellis's *On Thin Ice*, which describes the symbolic import and cultural significance of captive polar bears. We concluded with a consideration of Chanda Meek's "Putting the U.S. Polar Bear Debate into Context: The Disconnect between Old Policy and New Problems" (2011), which details some of the problems with current polar-bear management policies in arctic Alaska.[16]

Overall, I think that this unit of the course was the most engaging for the students. Most of them responded enthusiastically to the research assignment, because they had the opportunity to choose their own subject matter and pursue their own interests. I, too, enjoyed the assignment, for I learned about all sorts of endangered and extinct animals—from honeybees to woolly rhinoceroses. I especially appreciated that two of my students introduced me to de-extinction, the process of reviving a species through cloning or other biotechnologies. In their papers, they surveyed the available scientific literature, highlighting many of the logistical and ethical complications involved in de-extinction. Focusing specifically on the quasi-successful case of the Pyrenean ibex—a clone that lived just a few hours before succumbing to birth defects in its lungs—both concluded that extinct animals should remain extinct.

As their final assignment of the semester, students produced a portfolio composed of their three most accomplished papers and a reflection letter. They wrote meaningfully about the trajectory of the semester, highlighting their favorite assignments (the second and fifth) and activities (peer reviews and sentence workshops). Many students indicated that they appreciated all the feedback they received from me on the various drafts of their papers, and others gave me some good ideas for future improvements. Given UAF's fiscal climate, though, I am not sure that I will be able to implement many of their suggestions. With an increased teaching load and fewer university resources, I know that I will not be able to continue to assign five papers per semester and provide students with adequate feedback. Simply put, I am working at an unsustainable pace, and I will soon have to enact so-called efficiency measures of my own. Most likely, this reality will preclude creating new activities and implementing substantial changes to my course design.

Unsustainable Alaska: The Future

Since spring 2015, oil prices have not improved, and neither has Alaska's budget outlook. Although many Alaskans have come to accept the necessity of enacting taxes, the state's Republican legislators have refused to consider this option. In March 2016, Pete Kelly, cochair of the Senate Finance Committee, explained his logic thus: "I'm not getting into the tax business while I know government is still too big." Committed to this ideology, legislators continue to slash funding for state systems, services, and programs. The state House subcommittee with oversight of the University of Alaska's finances even briefly entertained Representative Tammie Wilson's proposal to cut funding for all non-teaching-related activities at the university, including scientific research and community outreach. While that proposal was rejected, other funding-reduction suggestions were not. So at present, UAF is preparing for a $75 million shortfall for 2016–2017, enacting even more stringent spending cuts, furloughs, and layoffs.[17] With no relief in sight, I suspect that I will continue teaching about sustainability in an unsustainable environment for many years to come.

NOTES

1. U.S. Department of the Interior, National Park Service, *Alaska Region Climate Change Response Strategy 2010–2014*, https://www.nps.gov/akso/docs/AKCCRS.pdf, 4; Ned Rozell, "Hybrid Grizzly-Polar Bear a Curiosity," *Fairbanks Daily News-Miner*, April 7, 2013; Weston Morrow, "UAF Researchers Show Arctic Leaking Massive Amounts of Methane," *Fairbanks Daily News-Miner*, November 27, 2013; Yereth Rosen, "Biologists Spot Huge

Gathering of Walruses on Beach Near Point Lay," *Alaska Dispatch News*, September 30, 2014; Rosen, "Some Alaska Animals May Benefit from Climate Change While Others Suffer, Studies Say," *Alaska Dispatch News*, March 28, 2015; "Study: Alaska's Melting Glaciers Key to Sea Level Rise," *Fairbanks Daily News-Miner*, June 17, 2015; Rachel D'Oro, "New USGS Report: Coastal Erosion Threatens Northern Alaska," *Fairbanks Daily News-Miner*, July 1, 2015; Robin Wood, "Alaska Wildfires Still Growing, Now Fifth Worst Year on Record," *Fairbanks Daily News-Miner*, July 13, 2015; Dorothy Chomicz, "Climate Stress Sets Fairbanks' White Spruce Up for Failure," *Fairbanks Daily News-Miner*, July 28, 2015.

2. Daniel J. Rohlf, "Six Biological Reasons Why the Endangered Species Act Doesn't Work—and What to Do About It," *Conservation Biology* 5, no. 3 (September 1991): 281; Stephen R. Kellert, "The Biological Basis for Human Values of Nature," in *The Biophilia Hypothesis*, ed. Stephen R. Kellert and Edward O. Wilson (Washington, DC: Island Press, 1995), 66; Richard Ellis, *On Thin Ice: The Changing World of the Polar Bear* (New York: Vintage, 2010), 257; Paul Ehrlich and Anne Ehrlich, *Extinction: The Causes and Consequences of the Disappearance of Species* (New York: Random House, 1981), 6; Paul Ehrlich, *The Population Bomb* (New York: Ballantine Books, 1968), 50–77. *The Population Bomb* imagines three disastrous scenarios for the future, involving thermonuclear war, global pandemic, and overpopulation. Although he explains, "these are just possibilities, not predictions," he emphasizes that these are "the kinds of events that might occur in the next few decades" (49).
3. "Facts and Figures," Alaska Oil and Gas Association, http://www.aoga.org/facts-and-figures; "At Long Last, a State Budget: Legislature Reaches Budget Compromise, Avoids Government Shutdown," *Fairbanks Daily News-Miner*, June 14, 2015.
4. "Budget and Planning," University of Alaska, Fairbanks, http://www.uaf.edu/.
5. Joe Paskvan, "Alaskans Lose with Oil Tax Reform Measures," *Fairbanks Daily News-Miner*, March 20, 2013; Jim Jansen, "Alaska Must Be Competitive," *Fairbanks Daily News-Miner*, April 7, 2013; Casey Grove, "Interior Secretary Jewell Meets Alaska Legislators on Northwest Tour," *Fairbanks Daily News-Miner*, February 17, 2015; Lisa Murkowski, "President Should Listen during His Alaska Visit," *Fairbanks Daily News-Miner*, August 28, 2015.
6. "Department of Mining and Geological Engineering," University of Alaska, Fairbanks, http://cem.uaf.edu/mingeo.aspx; "Department of Petroleum Engineering," University of Alaska Fairbanks, http://cem.uaf.edu/pete.aspx.
7. "Fall 2014 UAF Fact Sheet," University of Alaska Fairbanks, http://www.uaf.edu/pair/factsheet/; "University of Alaska–Fairbanks," *US News & World Report*, http://colleges.usnews.rankingsandreviews.com/best-colleges/alaska-fairbanks-1063.

8. David A. Gruenewald, "The Best of Both Worlds: A Critical Pedagogy of Place," *Educational Researcher* 32, no. 4 (May 2003): 3.
9. Georges Cuvier, "Memoir on the Species of Elephants, Both Living and Fossil," in *Georges Cuvier, Fossil Bones, and Geological Catastrophes: New Translations and Interpretations of the Primary Texts*, ed. and trans. Martin J. S. Rudwick (Chicago: University of Chicago Press, 1997), 24; Charles Lyell, *Principles of Geology* (London: J. Murray, 1832), 2:142. The allusion comes from *Henry V*, act 5, scene 2.
10. Henry Tukeman, "The Killing of the Mammoth," *McClure's Magazine*, October 1899, 505–514; Frederic A. Lucas, "The Truth about the Mammoth," *McClure's Magazine*, February 1900, 349–355.
11. Tukeman, "Killing," 512, 508–509.
12. Rohlf, "Six Biological Reasons," 281.
13. Richard Nelson, "Searching for the Lost Arrow: Physical and Spiritual Equality in the Hunter's World," in *The Biophilia Hypothesis*, ed. Stephen R. Kellert and Edward O. Wilson (Washington, DC: Island Press, 1995), 226.
14. Kellert, "Biological Basis for Human Values," 66; Nelson, "Searching for the Lost Arrow," 226.
15. The proposed Pebble Mine is a copper mine to be located in the Bristol Bay region of southwestern Alaska; the proposed Susitna Dam is a hydroelectric project to be located on the Susitna River in south central Alaska.

    For discussion of endangered wildlife, see U.S. Department of the Interior, U.S. Fish & Wildlife Service, "Questions and Answers about Bald Eagles Recovery and Delisting," last modified October 2012, http://fws.gov/midwest/Eagle/recovery/qandas.html; Tom Kenworthy, "Goose to Be Taken off Endangered List," *USA Today*, June 19, 2001.

    The pipeline has suffered periodic sabotage attempts and mechanical failures, at least two of which have resulted in significant, but not disastrous, ruptures. Admittedly, the oil spilled by the grounding of the *Exxon Valdez* on the Bligh Reef in Prince William Sound came from the Pipeline, and many Alaskans still harbor antipathy about this devastating disaster. Remarkably though, the Pipeline itself has escaped censure. For many Alaskans, it is a marvel of human engineering and a popular tourist attraction. See "The Alaska Pipeline," *American Experience*, dir. Mark Davis, aired 2006, on PBS, http://www.pbs.org/wgbh/amex/pipeline/peopleevents/e_environment.html.
16. Ellis, *On Thin Ice*, 225–257; Samuel Hearne, *A Journey from Prince of Wales's Fort in Hudson's Bay to the Northern Ocean*, ed. Richard Glover (Toronto: Macmillan, 1958); Chanda L. Meek, "Putting the US Polar Bear Debate into Context: The Disconnect between Old Policy and New Problems," in "The Human Dimensions of Northern Marine

Mammal Management in a Time of Rapid Change," ed. Chanda L. Meek and Amy Lauren Lovecraft, special issue, *Marine Policy* 35, no. 4 (July 2011): 430–439; Frederick Schwatka, *Nimrod of the North; or, Hunting and Fishing Adventures in the Arctic Regions* (New York: Cassell, 1885).

17. Nathaniel Herz, "New Rasmuson Poll Says Alaskans Want Action on Budget Crisis," *Alaska Dispatch News*, August 13, 2015; Herz, "Alaska Senate Leaders: We're Not Getting into 'the Tax Business,'" *Alaska Dispatch News*, March 15, 2016; Dermot Cole, "Wilson Would Slash up to 1,000 UA Jobs with Disastrous Plan," *Alaska Dispatch News*, February 20, 2016; Tegan Hanlon, "UAF Won't Get a New Chancellor as University Officials Consider Restructure," *Alaska Dispatch News*, May 12, 2016.

BIBLIOGRAPHY

"The Alaska Pipeline." *American Experience*. Directed by Mark Davis. Aired 2006, on PBS. Http://www.pbs.org/wgbh/amex/pipeline/peopleevents/e_environment.html.

Cuvier, Georges. "Memoir on the Species of Elephants, Both Living and Fossil." In *Georges Cuvier, Fossil Bones, and Geological Catastrophes: New Translations and Interpretations of the Primary Texts*, edited and translated by Martin J. S. Rudwick, 18–24. Chicago: University of Chicago Press, 1997.

Ehrlich, Paul. *The Population Bomb*. New York: Ballantine Books, 1968.

Ehrlich, Paul, and Anne Ehrlich. *Extinction: The Causes and Consequences of the Disappearance of Species*. New York: Random House, 1981.

Ellis, Richard. *On Thin Ice: The Changing World of the Polar Bear*. New York: Vintage, 2010.

Gruenewald, David A. "The Best of Both Worlds: A Critical Pedagogy of Place." *Educational Researcher* 32, no. 4 (May 2003): 3–12.

Hearne, Samuel. *A Journey from Prince of Wales's Fort in Hudson's Bay to the Northern Ocean*. Edited by Richard Glover. Toronto: Macmillan, 1958.

Kellert, Stephen R. "The Biological Basis for Human Values of Nature." In *The Biophilia Hypothesis*, edited by Stephen R. Kellert and Edward O. Wilson, 42–72. Washington, DC: Island Press, 1995.

Lucas, Frederic A. "The Truth about the Mammoth." *McClure's Magazine*, February 1900, 349–355.

Lyell, Charles. *Principles of Geology*. Vol. 2. London: J. Murray, 1832.

Meek, Chanda L. "Putting the US Polar Bear Debate into Context: The Disconnect between Old Policy and New Problems." In "The Human Dimensions of Northern Marine Mammal Management in a Time of Rapid Change," edited by Chanda L. Meek and Amy Lauren

Lovecraft. Special issue, *Marine Policy* 35, no. 4 (July 2011): 430–439.

Nelson, Richard. "Searching for the Lost Arrow: Physical and Spiritual Equality in the Hunter's World." In Kellert and Wilson, *The Biophilia Hypothesis*, 201–228.

Rohlf, Daniel J. "Six Biological Reasons Why the Endangered Species Act Doesn't Work—and What to Do About It." *Conservation Biology* 5, no. 3 (September 1991): 273–282.

Schwatka, Frederick. *Nimrod of the North; or, Hunting and Fishing Adventures in the Arctic Regions*. New York: Cassell, 1885.

Tukeman, Henry. "The Killing of the Mammoth." *McClure's Magazine*, October 1899, 505–514.

U.S. Department of the Interior. National Park Service. *Alaska Region Climate Change Response Strategy 2010–2014*. Anchorage, AK: National Park Service, 2010. Https://www.nps.gov/akso/docs/AKCCRS.pdf.

PART 2.

Rethinking What We Do, Remaking Curricular Ecologies

Letting the Sheets of Memory Blow on the Line: Phantom Limbs, World-Ends, and the Unremembered

Derek Owens

> that there will come so many terrors to us
> and to our measured hours.
> Or that, after many mornings, many days,
> we would come to love
> this waking life enough to dread its loss.
>
> —K. A. Hays, *Dear Apocalypse*

I suppose we're all haunted now on many levels. Remembrance by its nature seems a form of haunting. As is consciousness itself. During the day we recall faint wisps of last night's dreams, which have mostly evaporated into air, and strain to pluck the cottony scenes still floating about, like dandelion seeds, barely in reach.

The phrase in my title comes from Anne Carson's *Nox* (2010), her patchwork memoir about a brother of hers who had died without her knowing. Another ghost story.

Remembrance, we now know, is not the act of literally bringing back that which had been held in storage. Because what we recollect is ever a construction, remembrance is always architectural: forging, building, splicing. Remembrance is collage, assemblage. The gluing, mashing, mixing, sampling of ephemera, stand-ins for our concepts of "self," "history," "family."

But remembrance, to me, is also to be haunted by the (enormous, and unreachable) gaps in the map that can never be filled. The dreams and histories of the unremembered, and which can never be known (since to know something is to remember it), they vastly, infinitely, outnumber the remembered. Memory, then, a kind of mourning, a longing, a sad lust for that which can never come back—and yet we know it is (was) out "there," somewhere. On the tip of one's tongue, but a longing for a taste that can never be recaptured. "Saturated with unrequited longing."[1] *Presque vu.*

Remembrance is to be in the company of phantom limbs. Invisible legs and arms that are always there, even though gone; they itch, and they can be scratched. Dead grandmothers and pets we talk to throughout the course of our day, who sometimes talk back. The palimpsest of a city street occasionally revealing hints of its prior manifestations. Ghost relics, like the avocado, and other weird fruit engineered to be eaten by giant ground sloths, glyptodonts, and megafauna that vanished thousands of years ago.[2] All those echoes and ghost relics.

But in addition to this chorus of phantom limbs we carry inside us, our sense of remembrance, today, culturally and collectively—"worldwide"—might be said to be haunted too by the *unremembered*. The unremembered is an absence of what once was, and we can only know it via the abstract, the statistical—a measuring of that for which we have no knowledge. (Parallels, here, with particle physics and quantum mechanics—tracing what cannot be seen via the tracing of tracings.) Who, for example, can know how many thousands of unknown, unnamed, undiscovered insect, plant, and animal species have vanished in Indonesia so Dunkin' Donuts could get the palm oil necessary to make its more than one hundred (Who knew? Check their website!) different varieties of donuts?

Polar bears, gorillas, elephants, and other charismatic megafauna whose days are numbered: they're all well on their way to becoming phantom limbs. But as for the ever unnamed, unseen, obliterated-before-setting-foot-inside-our-consciousness—such species are not phantom limbs but rather the unremembered. What does it mean to mourn phantom limbs that have already become, or are becoming, extinct? What does it mean to contemplate the unremembered—to mourn what we can never know we are mourning? And how does one mourn in the midst of a culture that finds it almost impossible to recognize the value of what has been lost?[3]

"Let the sheets of memory blow on the line." But sheets to climb under at night, in a failed attempt to cover ourselves with what is gone for good? Or sheets of paper to write upon, logging the endangered and the soon to be disappeared (as in Julianna Spahr's "Unnamed Dragonfly Species")? How to "come to grips" with all we've let slip through our hands.

But we can't come to grips with what we have no memory of ever holding in our minds. The unremembered remains unrendered. Forever. An inevitable byproduct of this is realizing how incomplete, and disintegrating, are those selves we identify as our own. Selves en route to becoming somebody else's phantom limb, and then, eventually, unremembered as well. ("What happens to names when time stops? Answer: Nothing happens: There is no when."[4])

Telling Trauma

"Trauma inevitably brings loss," writes Judith Herman; "traumatic losses rupture the ordinary sequence of generations and defy the ordinary social conventions of bereavement. The telling of the trauma story thus inevitably plunges the survivor into profound grief. Since so many of the losses are invisible or unrecognized, the customary rituals of mourning provide little consolation."[5] But if that's so—if our customary rituals are inadequate—speaking of, and talking back to, trauma is still essential for the healing that can come from testifying.

But what rituals and procedures are available for mourning mass die-offs and extinctions taking place at an incomprehensible rate, on a scale none of us can begin to imagine? This is not about focusing on environmental collapse elsewhere on the "planet Earth." "We can't mourn for the environment because we are so deeply attached to it—we *are* it," Timothy Morton writes. "We start by thinking that we can 'save' something called 'the world' 'over there,' but end up realizing that we ourselves are implicated." As a result, he calls for "preserv[ing] the dark, depressive quality of life in the shadow of ecological catastrophe," of acknowledging art as "grief-work."[6]

We know considerably more now about strategies for helping survivors of trauma than we did a generation ago. (Herman's insights might seem obvious to us now, but let's not forget that it was not until the 1970s that the term "child abuse" first appeared.) But our ability to comprehend, let alone respond to, the ongoing and innumerable extinctions, ecological injustices, and sufferings brought on by a colossal failure to live in accordance with an ecological ethics, is still in its infancy.

This is not to equate the more abstract "trauma" of contemplating global die-offs and irreversible ecological collapse in the Anthropocene with the all-too-visceral trauma that survivors of rape, abuse, violent crime, and horrible tragedy live through daily. *That* kind of trauma, of course, can generate such psychic and physical damage as to destroy one's very core. (Imagining polar bear extinction may not lead one to commit suicide, but experiencing child abuse, incest, enslavement, etc., certainly can.) Which is part of the problem. The depths of what we are losing and what we are in for cannot be overstated. And yet to call it "trauma" might feel a little . . . overstated? Like many, I've seen firsthand what post-traumatic stress disorder can do to the survivor of child abuse (*Memory's Wake*).[7] And my students often write about the traumas of sexual assault, gang violence, poverty, addiction, abusive partners, incarceration, depression, and suicide. In comparison, contemplating worldwide mass extinction can almost be seen as a kind of thought experiment, an intellectual exercise in imagining "the future without." Perhaps we are in the early stages of experiencing not so much post-traumatic stress disorder but *pre*-traumatic stress disorder—a pervasive undercurrent of anxiety, dis-ease, and melancholy from anticipating, on some level, even (especially) unconsciously, that we have entered the end times.

Herman writes, "Having come to terms with the traumatic past, the survivor faces the task of creating a future."[8] But also: coming to terms with the traumatic future, the survivor faces the task of creating meaning in the present. How to tell the story about our failure to tell this story.

The Awful Red Delicious Apple

Slavoj Žižek turns to Elisabeth Kübler-Ross to explain how we're all dealing with the "forthcoming apocalypse," responding with variations of the five stages of grief (denial, anger, bargaining, depression, acceptance). To read those who write to us from the "acceptance" stage is a sobering experience. James Lovelock:

> Even if we had time, and we do not, to change our genes to make us act with love and live lightly on the Earth, it would not work. We are what we are because natural selection has made us the toughest predator the world has ever seen. . . . It

> is as absurd to expect us to change ourselves as it would be to expect crocodiles or sharks to become through some great act of will, vegetarian.[9]

(Or, more to the point, vegan.)

And Stephen Meyer's surgically precise assessment of our monolithic cultural failure is worth quoting at length:

> Over the next 100 years or so as many as half of the Earth's species, representing a quarter of the planet's genetic stock, will functionally if not completely disappear. The land and the oceans will continue to teem with life, but it will be a peculiarly homogenized assemblage of organisms unnaturally selected for their compatibility with one fundamental force: us. Nothing—not national or international laws, global bioreserves, local sustainability schemes, or even "wildlands" fantasies—can change the current course. The broad path for biological evolution is now set for the next several million years. And in this sense the extinction crisis—the race to save the composition, structure, and organization of biodiversity as it exists today—is over, and we have lost.

His conclusion:

> Of course, the end of the wild does not mean a barren world. There will continue to be plenty of life covering the globe. There will be birds, mammals, and insects—lots of insects. Life will just be different: much less diverse, much less exotic, much more predictable, and much less able to capture the awe and wonder of the human spirit. Ecosystems will organize around a human motif, the wild will give way to the predictable, the common, the usual. Everyone will enjoy English house sparrows; no one will enjoy wood thrushes.
>
> We have lost the wild for now. Perhaps in five or ten million years it will return.[10]

Twenty years earlier, David Ehrenfeld expressed a similar sentiment in "Life in the Next Millennium: Who Will Be Left in the Earth's Community?" In his version, we will have house sparrows, gray squirrels, possums, Norway rats, pigeons, cockroaches, feral house cats, ailanthus, and phragmite. There will be no more polar bears, of course, or tropical forests, and the temperate forests will be on their way out. The occasional rhino might still be found in a zoo here and there; some orchids might be preserved in botanical gardens, along with "poorly representative collections in seed banks of what is left of the human agricultural heritage: a few varieties of African upland rice, more varieties of wheat, a smattering of beans. The fruit trees will fare even worse, although I suppose we will have the awful Red Delicious apple around forever, as a perpetual reminder of our sins."[11]

Responses to World-Ending

What does one *do* with this kind of thinking? Sign up for some "ecotour spectacle" so as to witness living ghosts—to "experience" firsthand what shall soon be outside experience forever? Peter Lamborn Wilson contends: "All authentic bits of surviving sadness suddenly become simulacra of themselves—soon no doubt we'll have Sadness Tourism (in fact all tourism is a kind of mourning). Pills to evoke a trace of genuine melancholia."[12]

Those preoccupied with world-ending offer up several strategies.

Document the phenomenon of living in a "culture of denial." Ross Gelbspan, one of our best-known journalists on climate change, is interviewed in the documentary film *Everything's Cool*. He speaks despairingly of how he no longer sees any point in trying to make arguments that have been made and proved so many times already: "What is the existential response in an age of collapse? I mean, we're talking about a real change in the evolution of our species and our civilization; we're talking about a really big deal change. How is an intelligent, honest person supposed to relate to that?" He then tells the camera he has no more incentive to try and persuade readers that climate change is happening and that it's due to our actions. And so the only book he has left in him is one where he reflects on what it means to live in a culture in denial. (As he says this, his wife nudges him to stop talking, shaking her head.)[13]

Morton urges us to acknowledge that the end of the world has already occurred:

> We can be uncannily precise about the date on which the world ended. . . . It was April 1784, when James Watt patented the steam engine, an act that commenced the depositing of carbon in Earth's crust—namely, the inception of humanity as a geophysical force on a planetary scale. Since for something to happen it often needs to happen twice, the world also ended in 1945, in Trinity, New Mexico, where the Manhattan Project tested the Gadget, the first of the atom bombs, and later that year when two nuclear bombs were dropped on Hiroshima and Nagasaki.

He continues, later: "The very feeling of wondering whether the catastrophe will begin soon is a symptom of its already having begun." For Morton, a sane response is through art, but only if art is heretofore understood as "grief-work."[14]

Žižek, in *Living in the End Times*, in a section titled "Welcome to the Anthropocene," cites Ed Ayres, who suggests that the default for coping with threats like the ones we face now is the willful cultivation of ignorance: "[A] general pattern of behavior among threatened human societies is to become more blinkered, rather than more focused on the crisis, as they fail."[15]

So, two possible paths here: investigate, unpack, stare in the face the reality of our failures, documenting and acknowledging them. Perhaps, as Morton might have it, recast that brutal honesty as a kind of grief-art. Or, if this is too hard (and for most it is; how could it not be?), then accept our inability to become unblinkered. Not a hedonistic embrace so much as unapologetically seeking out and valuing joy in what's here now—even, especially, if it won't be here tomorrow.

Ian Pindar and Paul Sutton, in their introduction to Felix Guattari's *The Three Ecologies*, advocate a rejection of religion and afterlife fantasies that pull us away from a 24/7 contemplation of life on earth: "Our species remains vulnerable. In order for it to survive, the twenty-first century must be atheist in the best sense: a positive disbelief in God, concerned only with, respectful of, terrestrial life. It will require the development of an immanent, materialist ethics, coupled with an atheist awareness of finitude, of the mortality of the species, the planet and the entire universe, and not an illusory belief in immortality, which is only a misplaced contempt for life." In order to correct this "misplaced contempt," Guattari calls for "the promotion of innovatory practices, the expansion of alternative experiences centered around a respect for singularity"; "individuals must become both more united and increasingly different." Of course, such (admittedly vague) wishful thinking is in stark contrast with recent polls showing that Americans are increasingly dismissive of realities like climate change and evolution.[16]

Surely one of the impulses behind the popularity of OOO (object-oriented ontology) and the new materiality must be an awakening, a discovery, that one can come to love *things*. For so long "things" were equated with rampant consumer culture—whoever dies with the most toys wins. But things—objects—now refer to world(s) "outside" the "human": tulip trees, vertebrae, garnets, bodies, adolescent cats, verbena, collapsed barns, bees, sand, firewood, trains, board games, crayons . . . (the lists can and should go on for pages). I think this is what many visual artists feel on a rather primal level, a love that comes from looking intensely at the stuff of the world. Manipulating things with one's hands. To love things, and have empathy for them, is to reconceptualize ourselves as (happily) smaller than we realize, perpetually interlaced with a world of objects, from the bacteria in our guts to the software used to engineer the chairs we're sitting on. To train oneself to see (and hear, smell, taste) one's surrounds as a world of ready-mades, found art.

Catriona Mortimer-Sandilands locates a preservational impulse via a rethinking of *melancholia*:

> At the heart of the modern age is indeed a core of grief—but . . . that "core" is more accurately conceived as a condition of *melancholia*, a state of suspended mourning in which the object of loss is very real but psychically "ungrievable" within the confines of a society that cannot acknowledge nonhuman beings, natural environments, and ecological processes as appropriate objects for genuine grief.
>
> Melancholia, here, is not a failed or inadequate mourning. Rather, it is a form of socially located embodied memory in which the loss of the beloved constitutes the self, the persistence of which identification acts as an ongoing psychic reminder of the fact of death in the midst of creation. In a context in which there are no adequate cultural relations to acknowledge death, melancholia is a form of preservation of life—a life, unlike the one offered for sale in ecotourist spectacle, that is already gone, but whose ghost propels a *changed* understanding of the present.

She calls for "*queer* arts of memory, in which environmental histories and knowledges are rewritten as part of a memorial project."[17]

Echoing this, Nicole Seymour advocates for "queer empathy" where

> in our cynical "post"-everything age, individuals might learn to care about those persons and entities to whom they have no direct ties . . . [an] empathetic, politicized advocacy for the nonhuman natural world . . . a specifically queer ecological ethic of care: a care not rooted in stable or essentialized identity categories, a care that is not just a means of solving human-specific problems, a care that does not operate out of expectation for recompense.

She concludes: "The kind of empathy that environmentalism at large calls for so urgently right now is by definition queer, even when not directly linked to (homo) sexuality or sexual issues: one must care for nameless, faceless future beings, including non-humans, to which one has no domestic, familial, or financial ties."[18]

And then there are those who have used their art to advocate leaving entirely. In a recording made several years after the murders of Malcolm X and Martin Luther King Jr., June Tyson, singing in Sun Ra's Intergalactic Research Arkestra, conjures an alternative world in her singsongy birdcall: "black myth . . . of the living dream . . . this strange dream . . . in my dream world . . . a world, a world . . . black myth . . . a world . . . dream . . . world . . . black myth, a world . . . a world . . ." Sun Ra closes the piece by berating the audience with his chant: "It's after the end of the world—don't you know that yet?" Other musicians take a more nihilistic view (Tool's "Some say the end is near / Some say we'll see Armageddon soon / I certainly hope we will because / I sure could use a vacation from this . . ."); some prefer to simply dance (Medeski, Martin & Wood's *End of the World Party* [*Just in Case*]).[19]

"Guessing the apocalypse" remains a tremendous source of entertainment for consumers of literature, film, and television. In a previous draft of this essay, I was tempted to list all of the novels, movies, and television shows since World War II depicting post-collapse or post-apocalypse scenarios and landscapes, but it more than quadrupled the length of this chapter. (*Wikipedia*'s list of apocalyptic fiction comes to more than fifty pages.[20]) Post-collapse and survivalist adventure tales are so popular in the young adult fiction industry as to almost be passé. When the movie *Armageddon* came out, McDonald's used the film in a marketing campaign, issuing coupons with ads depicting a soda, burger, and fries floating in space in front of a flaming asteroid hurtling toward earth, with captions reading "Get Astronomical Savings Before Time Runs Out" and "Get These Big Savings Before It's Too Late." It can be exciting, this end-of-the-world business—*fun*, even. I'm tempted to write that we need to fight this end-times eros but that would be disingenuous. (I sometimes enjoy a good apocalyptic film as much as the next armchair survivalist.) Maybe the challenge is how personally to absorb this shift, contemplating how to do so with tact and empathy. And reject, or at least question, the kitschier and more obligatory and predictable of responses.

Cultural Sustainability

To bring it back to the classroom: I'm interested, now, in imaging the implications of addressing cultural sustainability in the curriculum, motivated by Jon Hawkes's *The Fourth Pillar of Sustainability: Culture's Essential Role in Public Planning* (2001). The three pillars of sustainability have been understood as ecological sustainability, social sustainability, and economic sustainability. Hawkes argues that cultural sustainability is the equally important but often overlooked fourth pillar.

He starts with two definitions of culture:

> The social production and transmission of identities, meanings, knowledge, beliefs, values, aspirations, memories, purposes, attitudes and understanding; the "way of life" of a particular set of humans: customs, faiths and conventions, codes of manners, dress, cuisine, language, arts, science, technology, religion and rituals; norms and regulations of behavior, traditions and institutions.

He argues that "a sustainable society depends upon a sustainable culture. If a society's culture disintegrates, so will everything else. . . . Vitality is the single most important characteristic of a sustainable culture." Hawkes later cites an imperative taken from the Intergovernmental Conference on Cultural Policies for Development, Stockholm: "Sustainable development and the flourishing of culture are interdependent."[21]

Among the characteristics he associates with this view of healthy, sustainable culture are *well-being, diversity, creativity and innovation*, and *a sense of place.* Regarding "diversity," Hawkes writes that

> Just as biodiversity is an essential component of ecological sustainability, so is cultural diversity essential to social sustainability. Diverse values should not be respected just because we are tolerant folk, but because we must have a pool of diverse perspectives in order to survive, to adapt to changing conditions, to embrace the future.

Culture, in other words, is "the expression and manifestation of what it means to be human."[22]

All of this is especially appealing to me given where I teach. My university now boasts the second most diverse student body of U.S. colleges. We're located in the most culturally diverse place in the Western Hemisphere, the center of Queens, New York. (Some have estimated that this might be the most ethnically, racially, linguistically, spiritually diverse population on the planet.) And while it's a far cry from the fairly traditional general-education curriculum we currently have at the university, I can't help imagining how amazing it would be if we promoted a core curriculum built around students researching, testifying about, and archiving their local cultures via a range of methodological and disciplinary lenses.

In the spirit of a biodiverse curriculum, students wouldn't just explore their local cultural realities in conventional forms. To require everyone to write, say, some traditional mode of "research paper" as the means by which their insights are presented would be hypocritical. To value a diversity of cultures yet mandate the expression of that diversity via some prescribed mode of discourse would undermine the enterprise. It would also be disingenuous since cultural diversity includes rhetorical, textual, poetic, artistic, translinguistic diversity.

And so my utopian notion of not just the writing classroom, but the presence of writing throughout the entirety of a curriculum is one where forms proliferate. Autobiography, memoir, biography. Manifesto, rant, op-ed piece. Creative nonfiction, "research essay," the so-called persuasive essay. "Free" verse (projective or whatever), fiction, prose poem. Handmade books made out of string and cardboard. Digital portfolios. Photo essays, video, audio recordings. To invoke Jed Rasula: "As a defender of psychological diversity, I'd like to expunge the word 'poet' from our vocabulary. Wouldn't it be wonderful to commend a piece of writing without the special pleading of genre marching alongside like the change of guards at Buckingham Palace?"[23]

In other words, above all else, a reveling in forms. Not to privilege or disparage any form. But to delight in the sheer variety of them. (And how forms always beget new mutations.) To recognize that the privileging of a tiny few select forms amounts not just to fetishization but to a kind of rhetorical eugenics and cultural homogenization whenever we declare certain forms as intrinsically, "genetically" superior or inferior compared to others. This would mean being in solidarity with alternative Englishes, multiple vernaculars, code-meshing, home languages, and accented Englishes.

To seek out examples of the rhetorical and discursive biodiversity we have so successfully expunged from the academy, and bring into our classrooms multigenre, multimodal, multimedia experimental risk-taking. Not necessarily to be weirder-than-thou or experimental for the sake of being experimental (although I can imagine worse ways to spend one's time). But to delight in the variety of the cultures and lives our students expose us to. Ezra Pound's "make it new," cliché as it might be, continues to pack a lot of punch. What's easy to forget is how each of our students, by virtue of their wholly original, unique experiences and voices, make our courses new simply from having entered the room. In this vein, Jason Palmeri wraps up *Remixing Composition: A History of Multimodal Writing Pedagogy* (2012) thus:

> Ultimately, if we seek to value the diverse embodied knowledges of *all* students and teachers in the field of composition, we must embrace a capacious vision of

> multimodal pedagogy that includes both digital and nondigital forms of communication: live oratory and digital audio documentary; quilting and video gaming; paper-based scrapbooking and digital storytelling; protest chanting and activist video making.[24]

If our world (what poet Chuck Stein calls the Sad World) is destined to become more and more a world of homogeneity, of invasive species (all those grackles, all that phragmite) and sundry collapses, ecological and social and economic—if our children are to witness global extinctions and massive human die-offs from here on out—then might we teachers and parents and mentors and researchers insist upon, and commit ourselves to fashioning, wherever and however, preservational ethics and biodiverse imaginations? Within each other, our students, our curricula? Perhaps as a small gesture of defiant opposition to the wave upon wave of collapses and extinctions and narratives of loss occurring out there, hourly, outside the classroom walls (and within our psyches too). To consider the classroom as some temporary loft or heterotopia. A site of salvaging. A momentary retreat where we might see what happens if we focus on reflecting and remembering and preserving, if only for a short while. Our job being to listen to and, if we can, help our students and ourselves hold on to what is ever at risk of disappearing, while cultivating and sharing and delighting in (and sometimes mourning over) our own unique, singular, local stories—conveyed not in forms that we teachers privilege, but in forms chosen in part by the students, in accordance with their needs and motives. Teaching as shared testifying—noisy, chaotic, and utterly heterogeneous—while there is still memory to work from.

My little interest in cultural sustainability is just one tentative gesture among many. Documenting despair, expunging immortality, making grief-art, cultivating queer empathy, dancing before the spectacle of collapse ("a dance of attention"[25])—all viable responses. I think the real story here is that none of us knows what to do at this historical moment. In a time of hauntings and ghosts and phantom limbs—of anxieties born from the awareness that we are causing more extinctions than at any time since the Pleistocene—memory, trauma, violence all take on radically new auras. The rate of collapse outpaces our ability to get a handle on the situation. How we tell this impossible story, and situate ourselves within it, will increasingly become our defining struggle.

NOTES

1. Timothy Morton, *Ecology without Nature: Rethinking Environmental Aesthetics* (Cambridge, MA: Harvard University Press, 2007), 186.
2. Connie Barlow, *The Ghosts of Evolution: Nonsensical Fruit, Missing Partners, and Other Ecological Anachronisms* (New York: Basic Books, 2000).
3. Catriona Mortimer-Sandilands, "Melancholy Natures, Queer Ecologies," in *Queer Ecologies: Sex, Nature, Politics, Desire*, ed. Catriona Mortimer-Sandilands and Bruce Erickson (Bloomington: Indiana University Press, 2010), 333.
4. Susan Howe, *That This* (New York: New Directions, 2010), 20.
5. Judith Lewis Herman, *Trauma and Recovery* (New York: Basic Books, 1997), 188.
6. Morton, *Ecology without Nature*, 186–187; Timothy Morton, *Hyperobjects: Philosophy and Ecology after the End of the World* (Minneapolis: University of Minnesota Press, 2013), 197. Felix Guattari writes: "Now, more than ever, nature cannot be separated from culture; in order to comprehend the interactions between ecosystems, the mechanosphere and the social and individual Universes of reference, we must learn to think transversally. Just as monstrous and mutant algae invade the lagoon of Venice, so our television screens are populated, saturated, by 'degenerate' images and statements. . . . In the field of social ecology, men like Donald Trump are permitted to proliferate freely, like another species of algae, taking over entire districts of New York and Atlantic City; he 'redevelops' by raising rents, thereby driving out tens of thousands of poor families, most of whom are condemned to homelessness, becoming the equivalent of the dead fish of environmental ecology" (*The Three Ecologies* [London: Continuum, 2008], 29).
7. See Derek Owens, *Memory's Wake* (Brooklyn: Spuyten Duyvil, 2011).
8. Herman, *Trauma and Recovery*, 196.
9. Slavoj Žižek, *Living in the End Times* (London: Verso, 2010), xi; James Lovelock, *The Vanishing Face of Gaia: A Final Warning* (New York: Basic Books, 2009), 231.
10. Stephen M. Meyer, *The End of the Wild* (Cambridge, MA: MIT Press, 2006), 4–5, 90.
11. David Ehrenfeld, "Life in the Next Millennium: Who Will Be Left in Earth's Community?," in *The Last Extinction*, ed. Les Kaufman and Kenneth Mallory (Cambridge, MA: MIT Press, 1987), 178.
12. Mortimer-Sandilands, "Melancholy Natures," 333; Peter Lamborn Wilson, *Ec(o)logues* (Barrytown, NY: Station Hill Press, 2011), 65.
13. *Everything's Cool*, dir. Daniel Gold and Judith Helfand (Toxic Comedy Pictures and Lupine Films, 2006).
14. Morton, *Hyperobjects*, 7, 177, 197.
15. Žižek, *Living in the End Times*, 327.

16. Ian Pindar and Paul Sutto, introduction to Guattari, *The Three Ecologies*, 11; Guattari, *The Three Ecologies*, 39, 45; Michael Roppolo, "Americans Not Confident Big Bang or Evolution Is Real, Poll Shows," *CBS News*, April 21, 2014, http://www.cbsnews.com/news/americans-big-bang-evolution-ap-poll.
17. Mortimer-Sandilands, "Melancholy Natures," 333, 343.
18. Nicole Seymour, *Strange Natures: Futurity, Empathy, and the Queer Ecological Imagination* (Urbana: University of Illinois Press, 2013), 184–185.
19. Sun Ra and His Intergalactic Research Arkestra, *It's after the End of the World* (*Live at the Donaueschingen and Berlin Festivals*), recorded 1970, Universe, 2010, CD; Tool, Ænima, Sony Legacy, 1996, CD; Medeski, Martin & Wood, *End of the World Party* (*Just in Case*), Blue Note, 2004, CD.
20. "Apocalyptic and Post-Apocalyptic Fiction," *Wikipedia*, last modified May 9, 2016, https://en.wikipedia.org/wiki/Apocalyptic_and_post-apocalyptic_fiction.
21. Jon Hawkes, *The Fourth Pillar of Sustainability: Culture's Essential Role in Public Planning* (Melbourne: Common Ground, 2001), 3, 12.
22. Ibid., 14, 32. Expressions and manifestations are never static: as Roy Wagner writes in his book *Symbols That Stand for Themselves* (Chicago: University of Chicago Press, 1989), "The basic frames of culture are formed as large-scale tropes, essentially like myths . . . [and as such] live in a constant flux of continual re-creation" (129).
23. Jed Rasula, "Interview," in *Eco Language Reader*, ed. Brenda Iijima (Callicoon, NY: Nightboat, 2010), 143.
24. Jason Palmeri, *Remixing Composition: A History of Multimodal Writing Pedagogy* (Carbondale: Southern Illinois University Press, 2012), 160.
25. Erin Manning and Brian Massumi, *Thought in the Act: Passages in the Ecology of Experience* (Minneapolis: University of Minnesota Press, 2014), 4.

BIBLIOGRAPHY

Barlow, Connie. *The Ghosts of Evolution: Nonsensical Fruit, Missing Partners, and Other Ecological Anachronisms.* New York: Basic Books, 2000.

Carson, Anne. *Nox*. New York: New Directions, 2010.

Ehrenfeld, David. "Life in the Next Millennium: Who Will Be Left in Earth's Community?" In *The Last Extinction*, edited by Les Kaufman and Kenneth Mallory, 167–186. Cambridge, MA: MIT Press, 1987.

Everything's Cool. Directed by Daniel B. Gold and Judith Helfand. New York: Toxic Comedy Pictures and Lupine Films, 2006. DVD.

Guattari, Felix. *The Three Ecologies*. Translated by Ian Pindar and Paul Sutton. New York: Continuum, 2008.

Hawkes, Jon. *The Fourth Pillar of Sustainability: Culture's Essential Role in Public Planning*. Melbourne: Common Ground, 2001.

Herman, Judith Lewis. *Trauma and Recovery*. New York: Basic Books, 1997.

Howe, Susan. *That This*. New York: New Directions, 2010.

Lovelock, James. *The Vanishing Face of Gaia: A Final Warning*. New York: Basic Books, 2009.

Manning, Erin, and Brian Massumi. *Thought in the Act: Passages in the Ecology of Experience*. Minneapolis: University of Minnesota Press, 2014.

Medeski, Martin & Wood. *End of the World Party (Just in Case)*. Blue Note, 2004. CD.

Meyer, Stephen M. *The End of the Wild*. Cambridge, MA: MIT Press, 2006.

Mortimer-Sandilands, Catriona. "Melancholy Natures, Queer Ecologies." In *Queer Ecologies: Sex, Nature, Politics, Desire*, edited by Catriona Mortimer-Sandilands and Bruce Erickson, 331–358. Bloomington: Indiana University Press, 2010.

Morton, Timothy. *Ecology without Nature: Rethinking Environmental Aesthetics*. Cambridge, MA: Harvard University Press, 2007.

———. *Hyperobjects: Philosophy and Ecology after the End of the World*. Minneapolis: University of Minnesota Press, 2013.

Owens, Derek. *Memory's Wake*. Brooklyn: Spuyten Duyvil, 2011.

Palmeri, Jason. *Remixing Composition: A History of Multimodal Writing Pedagogy*. Carbondale: Southern Illinois University Press, 2012.

Rasula, Jed. "Interview." In *Eco Language Reader*, edited by Brenda Iijima, 125–145. Callicoon, NY: Nightboat, 2010.

Seymour, Nicole. *Strange Natures: Futurity, Empathy, and the Queer Ecological Imagination*. Urbana: University of Illinois Press, 2013.

Spahr, Juliana. *Well Then There Now*. Boston: Black Sparrow Press, 2011.

Sun Ra and His Intergalactic Research Arkestra. *It's after the End of the World (Live at the Donaueschingen and Berlin Festivals)*. Recorded 1970. Universe, 2010. CD.

Tool. Ænima. Sony Legacy, 1996. CD.

Wagner, Roy. *Symbols That Stand for Themselves*. Chicago: University of Chicago Press, 1989.

Wilson, Peter Lamborn. *Ec(o)logues*. Barrytown, NY: Station Hill Press, 2011.

Žižek, Slavoj. *Living in the End Times*. London: Verso, 2010.

Student Expectations, Disciplinary Boundaries, and Competing Narratives in a First-Year Sustainability Cohort

Corey Taylor, Richard House, and Mark Minster

HERMIONE: Pray you sit by us,
And tell 's a tale.
MAMILLIUS: Merry or sad shall 't be?
HERMIONE: As merry as you will.
MAMILLIUS: A sad tale's best for winter. I have one
Of sprites and goblins.
—William Shakespeare, *A Winter's Tale*

During summer 2010, the three of us—all English professors—joined colleagues from engineering, chemistry, mathematics, and student affairs to found the Home for Environmentally Responsible Engineering (HERE), a living-learning community that introduces sustainability to first-year undergraduates at Rose-Hulman Institute of Technology (RHIT) in Terre Haute, Indiana. While education for sustainability and living-learning communities are popular worldwide, HERE combined the two for the first time on our campus. Students who wish to join HERE complete a questionnaire and write a paragraph about their interest in sustainability. Those admitted to the cohort, up to forty-four students, live together in one residence hall, take a sequence of three classes, and participate in on- and off-campus activities with environmental emphases.

As the first program of its type at RHIT, HERE has made small but noticeable impacts on the curriculum and campus. Two extant first-year courses (Rhetoric and Composition and Introduction to Design) were overhauled to include sustainability learning outcomes, and a new course (Introduction to Sustainability) was created to survey the field. In 2014, led by HERE faculty, RHIT began offering a certificate in sustainability studies that incorporates three core courses, two electives, and one seminar, which requires students to have completed at least one sustainable engineering project in order to enroll. Moreover, our Department of Facilities Operations has implemented some student-designed projects, including new lighting in an academic building, a passive-solar greenhouse with a rainwater collection system, and a landscaping project that mitigates runoff. In addition, HERE students have turned a defunct Sustainability Club into a campus chapter of Engineers for a Sustainable World. The HERE program graduated its first members in May 2015, and to mark the occasion a brief article appeared in the RHIT alumni magazine.[1] For its efforts, HERE has earned two small grants, one internal for faculty education in Leadership in Energy and Environmental Design (LEED) practices, and one external from Procter & Gamble for student-designed sustainability projects.

Despite our modest successes, we and our students have often felt frustrated. The program continues to attract two dozen or so students in each year's entering class, but we have not seen the broad exposure other programs have, evidenced in part by institutional communication. Prior to May 2015, HERE had not been the subject of a stand-alone campus news article since 2011–12, when the program began. It was covered then in the local newspaper and mentioned in a journal article.[2] After these appearances, aside from our own scholarly activities, the HERE program had no institutional publicity or recognition until May 2015. Students continue to join the program, but we have never reached capacity, enrolling twenty-seven students in 2011–12, then thirty-four, thirty-three, nineteen, twenty-six, and in 2016–17, twenty students in a first-year class of 561.

We see the challenges of maintaining HERE's enrollment and of promoting enthusiasm about the program arising from the difficulty in getting traction for a discussion of sustainability in the face of a dominant, consumerist narrative about the economic returns on investment of undergraduate engineering education. (While this narrative increasingly pervades higher education, it has long held sway in engineering education; our own institution's case is merely the most visible to us.)

At the outset, we believed HERE would establish a path on which students might be motivated by different values, establishing their own alternative foundation for their undergraduate study and, eventually, their lives and careers. We hoped HERE would help engineering students become creative, flexible, critical thinkers able to map, even navigate, the deep waters of sustainability while coming to appreciate the interconnectedness of the world's natural and human systems. In our estimation, these hopes have been realized in so few of our students that we question our own value. Our intentions—to make sustainability more vital in the curriculum, to introduce students to the kind of problem-solving that engineers can bring to environmental challenges, to foster creative thinking and cross-disciplinary collaboration, and to nudge the too-common narrative of engineering education as narrow training for a specialized profession—seem to have been too lofty. Our efforts at fulfilling these intentions have achieved mixed results.

The environment of HERE is not uninhabitable, but the context in which we operate is suboptimal. No one at RHIT has ever said that the program cannot run or that sustainability does not belong in engineering education or practice. Various campus groups support HERE in word and in deed. Our context, therefore, is not *actively* unsustainable but *passively* so, the result of contextual factors with logistical and philosophical effects. Some of these effects manifest themselves actively, but their source remains passive, established by contexts out of our control. For instance, consider money and time: Our first yearly budget ($10,000) from the college came in our third year, and no HERE professor has ever been granted release time for planning or administering the program. We have applied for and received course-development funding, but actual collaborative teaching has always meant more work for the same pay.

More significant hindrances have arisen from philosophical differences. In recent years our institute, like many others, has emphasized "innovation" and "entrepreneurship," incorporating these buzzwords into our identity and curriculum. We have an Associate Dean of Innovation, a Visiting Professor of Practice of Entrepreneurship, and an Innovation Center; "innovation" is a value named in our strategic plan. (Sustainability, by contrast, is not mentioned.) "Innovation," "entrepreneurship," and other corporate jargon ("return on investment," "value added") dominate the twenty-first-century higher-education landscape.[3] To its credit, Rose-Hulman's conception of entrepreneurship has been expansive, insisting that an entrepreneur need not start a profit-driven enterprise. In fact, the school has funded some of our HERE course-development activity through the

Kern Engineering Entrepreneurship Network (KEEN) program, recognizing that KEEN's "three C's"—Curiosity, Making Connections, and Creating Value—inhere in sustainability education. Given the careers that engineering graduates traditionally pursue, though, the most frequent stories by far concern entrepreneurs who develop and commercialize high-tech products; in an era when Silicon Valley provides the dominant model for technology development, the search for "killer apps"—and their concomitant venture capital—looms large. We therefore remain skeptical that sustainability could ever be genuinely deployed in profit-driven contexts, even those with softened edges.

Thus, rather than being situated in a hostile environment, HERE resides in a passively unsustainable environment characterized by indifference. Even this indifference has negatively impacted the aims of our program and has led us to take action. The program has been supported, but mostly nominally; the program has permission to exist, but not encouragement to thrive. Coupled with the widespread ethos of practicality and the desire for profitable careers that our students are encouraged to embrace, our program's context does not allow it to flourish. HERE's passively unsustainable environment, we contend, hinges on three factors: student expectations about the curriculum; the disciplinary affiliation of the authors, which is often associated with so-called "soft skills"; and tension between sustainability and the practical ethos-entrepreneurship narrative. This third component, narrative, is the most important: How our institution communicates its conceptions about the education it provides contradicts and ultimately excludes HERE's intended narrative.

This tension mimics the tension between incompatible conceptions about the ultimate function of higher education. The narrative of higher education's function seems to have metamorphosed from preparation for citizenship into specialized, privatized, practical training for lucrative employment—a trend long in the making, but especially and understandably evident in the wake of the Great Recession. That students think critically, that they find happiness apart from consumption, that they become responsible citizens and empathetic thinkers, that they might question the worth of "innovation" and "entrepreneurship"—these qualities are drowned by the race for employment, seen either as impediments or irrelevancies. How can education for sustainability occur in a narrative context that systematically implies that it does not belong?

We want to stress that this chapter is not a complaint about our HERE students, who after all have chosen to study environmentally responsible engineering. Nor do we mean to disparage any person or group at our institution, or our institution as a whole, which does, according to our mission statement, "provide an atmosphere of individual attention and support" that is highly regarded. This is not a tale of woe, but a reflective commentary on our own program's history, contexts, and futures, with an eye toward helping similar programs at other institutions learn from our challenges and, hopefully, to overcome their own.

RHIT and the Roots of HERE

Rose-Hulman Institute of Technology is a four-year, coeducational, private college specializing in undergraduate engineering, science, and mathematics education. The college focuses on undergraduate programs, which include nineteen bachelor of science degrees, thirty-nine minors (including environmental chemistry and environmental engineering), and five certificate programs (including sustainability). Enrollment at the start of the 2016–17 academic year stood at 2,278 students, most of them undergraduates.

Despite its small size and mostly regional draw, RHIT's national reputation has grown over the past two decades. The college's most-cited accolade has come from *US News & World Report*, which for the last eighteen years has ranked it the top engineering college in the United States among those that do not offer a doctorate. Rose-Hulman frequently appears on lists that tout the economic advantages of earning a degree in a science, technology, engineering, or mathematics (STEM) field. In 2015 the Brookings Institution, for instance, named RHIT one of the top "value-added" colleges in the United States, where "value" equates to three metrics that center on earning money. Our alumni's earning power has been corroborated by *The Economist* and PayScale: in 2015, both of these entities deemed RHIT the number one school in Indiana—and twenty-third and thirteenth, respectively, in the United States—for achieving economic gain. As have many other schools, RHIT has told students to expect these outcomes, advertising a near-perfect job placement rate and high starting salaries for graduates; as of January 2017, fully 97 percent of the class of 2016 had found full-time employment, with an average salary of $68,929.[4] We do not begrudge anyone their income, and we are aware that skyrocketing tuition costs and debt loads are harsh realities for millions of college students, alumni, and their families. But the Brookings report and others like it underscore a single narrative about what college is for: to obtain earning power. Such assumptions powerfully affect the HERE program.

In 2007, when a group of faculty and staff were appointed to the Sustainability Team, Rose-Hulman joined the Association for the Advancement of Sustainability in Higher Education (AASHE), and the school's then-president signed the American College and University Presidents' Climate Commitment (ACUPCC).[5] Encouraged by these developments, we three were among the English faculty members who took the opportunity to revise an upper-division writing course, Technical and Professional Communication (TPC), and propose a faculty professional-development program modeled after the Ponderosa and Piedmont Projects. TPC is one of two writing courses that our undergraduates must take, and we revised it to include readings, activities, and assignments focused on sustainability and sustainable engineering. That version of the course culminated with a team project that identified and proposed solutions to on-campus environmental problems, in keeping with the trend of greening the campus.

For most students, TPC was their first and only exposure to sustainability, unless it was part of their research interests, senior design project, or career aspirations. These students took TPC mainly as juniors, when they were enrolled in advanced major courses. It eventually became clear that we had introduced third-year students to sustainability too late in their cognitive and professional development to impact their selection of courses or careers. Unsurprisingly, the element of sustainability received initial criticism on TPC course evaluations. Despite the negativity, we maintained that one sustainability-centered course late in the curriculum was better than none.

For some TPC faculty, it was also our first foray into teaching sustainability. How could we and other professors with similar interests inspire students to consider sustainable engineering as a career—let alone teach sustainability as a subject—if we were not trained formally? This question was an impetus for the second precursor to HERE, the Ouabache Project for Sustainability across the Curriculum. This internal grant proposal called for summer workshops for up to fifty RHIT faculty members from across departments. These workshops, run by sustainability experts over three summers, would help faculty incorporate sustainability principles and modules into existing courses, and in so doing integrate sustainability across the curriculum. While such cross-disciplinary faculty-development workshops in sustainability are not uncommon, funding for such an initiative at RHIT would have been a first. Academic administrators rejected the proposal, stating that students did not attend RHIT to learn about sustainability, even though that was precisely the problem the Ouabache Project sought to solve.[6]

Nevertheless, we knew that sustainability needed a presence early in the curriculum. At a summer faculty-development workshop in 2010, we cofounded HERE and wrote a grant proposal to the Fund for the Improvement of Post-Secondary Education (FIPSE). Though we did not receive a grant, our proposal was well received by some reviewers. That fall, our attempts at securing funding from engineering firms in Indianapolis ended similarly, largely because our appeals were unsupported by our development office. We quickly realized that the program would need to be facilitated with minimal or no funding.

Thus HERE is truly grassroots, created by faculty and staff from across the institute who wanted to incorporate sustainability into the curriculum by teaching it during students' first year. HERE does not belong to any department, which has meant that faculty retain control over the program, and yet we are largely on our own—acknowledged, certainly, by the institute, but not supported by it. This, which seems at times to be a sin of omission, or benign neglect, has impacted HERE's functionality. Our ground-level efforts have not been reinforced at the upper levels, despite the nominal institutional commitment to sustainability signaled in 2007. Such bottom-up or "micro-level" changes to sustainability education, among them fewer restrictions on their growth, less interference from administrative or outside entities, and a leaner, more efficient logistical model, are to be counted as benefits, according to Matthias Barth. HERE possesses some of these features, but they are not necessarily advantageous. Barth concedes that even if faculty and staff bring their energy and intellect to developing sustainable curricular initiatives, administrators must match faculty efforts for those initiatives to succeed: "A lasting implementation can hardly be achieved without commitment and support at this stage from the top."[7] HERE's free-agent nature is both its greatest strength and biggest weakness.

Decoding the Expectations of Eighteen-Year-Old Engineering Students

We welcomed the first HERE cohort in August 2011, certain that students would warm to the material even though it was untested. They had, after all, elected to join a sustainability program created just for them. Prior to the first HERE cohort's arrival, we often asked ourselves, "What do the students expect?" We know what most HERE students want less of, based on course evaluations: less traditionally humanistic content such as assigned readings and writing assignments. One comment, made in conversation with a student from the 2012 cohort, encapsulates a common student reaction to our teaching in the humanities classroom: "We don't do anything. We read and talk and think and do community service, but we don't do anything."

Reading, talking, and community service do not constitute *actions*, or things that are *done*, despite community service being a distinct type of action. These are among the main components of the living-learning community, and our students do not typically consider them work. The comment above laments the lack of *technical content* in the HERE program—use of technology constitutes action for many of our students. Chiefly technical interests and action-oriented dispositions do not inherently preclude thoughtfulness and a deep regard for the implications of actions or of the frameworks in which those actions have meaning. In engineering education, Donna Riley, for one, has long championed the particular power of engineering's bent toward action to improve human and ecological well-being alike:

> The engineer wants to help . . . , whether it is bringing clean water and sanitation to a community or developing new drugs, designing renewable energy solutions to address climate change, or connecting people with wireless networks. Engineers are known for our work ethic; we are committed to getting the job done and will slog through hours of grunt work to make it happen. We serve and serve well. The helping spirit and strong work ethic of engineers are important traits for engaging in social justice work.

More recently, philosopher Matthew B. Crawford has championed "doing" forms of mechanical and skilled manual labor as an antidote to solipsism and a tool for advanced introspection. Crawford's advocacy of the skilled trades and mechanical work remind us that engineering has often been the profession of choice for first-generation college students whose families have practiced these trades or operated farms. Engineers are intellectually formidable, but their hands-on practicality lends them a certain trustworthiness.[8]

Riley and Crawford define technical competencies very differently, but both emphasize the transformative possibilities of pairing them with rigorous study in the humanities and social sciences. (The two are most alike in fiercely resisting any promotion of "soft skills" toward the aim of advancement in a corporate career.)

Unsurprisingly, few first-year engineering students share this outlook. Students in previous cohorts have complained—in evaluations and in meetings with HERE faculty and staff—that the variant meanings of sustainability, the cultural and political contexts and conceptual frames of what they are learning, seem unnecessary for their future work in their majors and eventual careers. As a result, many of our students are excited about photovoltaics, rain gardens, microturbines, and other forms of what David Orr calls "technological sustainability," which turns to technical objects and "market solution[s]," whereas ecological sustainability is a more fundamental attempt to rethink values and cultural practices. While technological sustainability sees economic growth and economic self-interest as essential for sustainable development, ecological sustainability means accepting natural limits to human economy and to the scale of communities and cultures. HERE has attempted to broaden the view of what an engineering education can provide—for instance, by emphasizing Orr's call to "restore civic virtue, a high degree of ecological literacy, and ecological competence throughout the population"—but in so doing risks being considered irrelevant by the group that should be our most vocal champion.[9]

That students should define "relevance" narrowly is nothing new. Although many RHIT faculty and staff resist the notion that engineering education should be purely instrumental, many more engineering students expect just that, having been encouraged to see higher education as a commodity or a business transaction. They are recruited on the premise, the admissions pitch, that they will likely become highly paid professionals, implying that education is superfluous if it does not contribute to that goal.

Many of our students are disinterested in the contextualization required by Orr's vision of ecological sustainability, even when it comes to looking critically at their own communities. By failing to see relevance in the social scope and environmental impacts of the problems their gadgetry hopes to address, our students will be just as likely to come up with solutions that cause new problems as to solve old ones. If students persist in seeing both college and sustainability *only* as careerist pursuits, they will miss out on other dimensions of college and sustainability. Similarly, if sustainability remains tangential in higher education—especially in engineering, science, and mathematics—then when students become practitioners in those fields, they might be unable or unwilling to address the shared problems humanity faces, when they should be well suited to help solve those problems.

In RHIT, as in many engineering curricula, the first two years cover foundational mathematics and science, design skills, engineering fundamentals, and professional practices. The inherent complexity of sustainability principles often forces their relocation to later in the educational process, leading to a treatment that is too delayed to be meaningful, as we found while teaching TPC. Our living-learning community derives from the belief that if we reach students before they become ensconced in professional silos, and if they live and learn together, then we could build something valuable. We were genuinely surprised to learn that the first cohort was more interested in learning about robotics, automotive engineering, and particle physics than about renewable-energy initiatives in India, listening to guest lectures on sustainable workplace practices, and attending a conference on community environmental initiatives. Unsurprisingly, it has been the *doing* activities—energy audits, water sampling and testing, cocurricular design projects—that students have viewed favorably, but without reflection on underlying conceptual frameworks and cultural contexts, which students resist, we believe such doing is fruitless.

HERE has attempted to offer an alternative to technological sustainability—sustainability seen as a design constraint or merely a professional practice. We wanted the program to instill in our students what Kevin Warburton calls "deep learning," which involves moving beyond teaching facts to the creation of "an active, transformative process of learning that allows values to be lived out and debated, and permits a unification of theory and practice." But our attempts at deep learning have not caught on, likely because such learning does not square with students' expectations about engineering education. Students have expressed frustration at a perceived discrepancy between their professional goals and our curriculum, in which the goals and traditions of the liberal arts, interrogating and contextualizing arguments and values, preceded the practice and processes of engineering design. Others have seen challenges to their beliefs as threatening, and have even said that they should not be asked to approach engineering problems from historical, cultural, and rhetorical perspectives. Such frustrations have prompted us to rethink the program's goals of achieving meaningful education for sustainability during the first year, especially because students who view engineering education and sustainability as on-the-job matters miss out on the attitudinal, behavioral, and civic dimensions of sustainability that are integral to the transformation demanded by both higher education and sustainability.[10]

Curriculum Design, Redesign, and Redesign Again

Our attempts to fulfill students' expectations have led us to reshape the HERE curriculum every year. Of all the effects caused by our passively unsustainable environment, our curricular redesigns have been the most active and time-consuming. In the program, students complete four courses over three ten-week quarters: College and Life Skills, and Rhetoric and Composition in the fall; Sustainability and Its Global Contexts in the winter; and Introduction to Design, formerly in the spring. Because students felt they weren't "doing" enough, we delayed Introduction to Design until the spring.

To address student concerns, in 2014 we merged Introduction to Design, and Rhetoric and Composition into a two-quarter-long (fall and winter) sequence team-taught by English and Mechanical Engineering professors. Now, students could immediately begin designing solutions to on-campus problems. We moved Introduction to Sustainability to the spring so students could reflect on their work. Folding the requirements of a writing course into the design process allowed for more reflection and in-depth contextualization of the problems needing to be solved. So rather than decontextualizing engineering problem-solving as some of them had hoped, we doubled-down on showing how "problem-setting," to use Donald Schön's phrase, informs the processes of engineering design, of writing, and of education for sustainability: "In real-world practice, problems do not present themselves to the practitioner as givens. They must be constructed from the materials of problematic situations which are puzzling, troubling, and uncertain."[11] While we intended this redesign to be more realistic and intensive, students resisted the longer design process as well as the delayed sustainability content.

In 2015–2016, we revisited the linked courses sequence, to move from global sustainability in the fall, to regional sustainability in the winter, and campus sustainability in the spring. We are experimenting with problem-based learning (PBL), which appeals to the notion of engineers as problem solvers while concentrating on problems of unsustainability, as Steinemann observes: "Typically, students . . . have relatively little experience in solving ill-structured, open-ended problems—the type often faced in practice. This is especially true with sustainability problems, which require flexible, integrative, multidisciplinary problem-solving approaches, rather than singular solutions."[12] To assess our changes, three times per year students are completing a rubric based on the program's learning outcomes in order to chart at what levels they feel their learning is occurring. We want students to practice metacognition and reflection, and to take some responsibility for the meanings they derive from the HERE program.

While students are not formally involved in curriculum design, we take seriously and try to enact thoughtful suggestions. Faculty and staff are keenly aware of student frustrations, which themselves frustrate us, as do the changes we make due to student critiques. The week-to-week, year-to-year revisions have not yet allowed the curriculum to find its form as we apply one year's comments to the next cohort's experience. Nevertheless, we have been able to plan for and manage some of our students' frustrations, encouraging persistence and identifying on-campus areas of need for new projects. To appeal to potential HERE members we have begun—carefully—to adopt the language of the college's dominant narrative, promoting our living-learning community by claiming that it "adds value" to their college experience and provides evidence of the "innovation" desired by employers. We do not enjoy co-opting the language of the narrative that excludes us, and we recognize the dangers, but we do so to better communicate some of HERE's goals to our audience. Still, students persist in expressing disappointment with the parts of HERE they deem irrelevant—the ones in which we explore the social scope and environmental impacts of the problems their engineering work hopes to solve. We have begun to think that maybe it's not them, but us.

Humanists, Sustainability, and Disciplinary Boundaries

Sustainability is inherently multidisciplinary, equally adept at finding a home in STEM, humanistic, and social science fields and in professional and business settings. The faculty and staff of HERE are a multidisciplinary group, and have become even more so since 2011. We have added to our roster a civil engineer, a computer scientist, another mathematician, an environmental chemist, and staff members in enrollment management and student life. The influx of new members and perspectives has been instrumental to HERE's development. Everyone shares responsibility for attending meetings, generating ideas for the program, responding to student needs, and participating in the living (cocurricular) side of the program. Indeed, our best ideas come from our group meetings. And while we share administrative duties, a mechanical engineer served as the first director—a calculated move to grant HERE legitimacy, given our institution's identity and mechanical engineering's popularity.

Work behind the scenes on HERE may be widely distributed, but teaching is more limited due to the courses on offer. The mechanical engineering professor who directed HERE for its first five years taught Introduction to Design, but Rhetoric and Composition and Introduction to Sustainability are offered by liberal arts faculty. The authors have taught both courses, in all of their iterations, every year for five years, which means that the majority of teaching in an "engineering" program is done by English professors. HERE's multidisciplinarity—and our independence from an academic department—remains a boon and a burden. The burden is felt most acutely in the classroom, where the limits of multidisciplinarity are apparent and disciplinary boundaries are visible.

The second feature of our passively unsustainable environment, then, is the disjunction between the authors' disciplinary affiliation and our students' developing professional identities. Theoretically, sustainability seems to exemplify how engineers need humanists, social scientists, and natural scientists to help with identifying problems, and how scientists and humanists need engineers to sort through the complex products and processes that solve, create, and solve problems again.[13] But practically, each professor, workplace, and department is characterized by expertise in particular technologies, products, or processes. Environmental responsibilities are often regarded as the purview of those with specialties such as geotechnical or water resource engineering, rather than as a set of skills, values, and habits for all college graduates. Without sufficient disciplinarity, sustainability can seem to have too little content; with too much disciplinarity, students and faculty alike miss the opportunity to address real-world problems with the range of skills and thinking they require.

Scholars and practitioners of sustainability acknowledge its multidisciplinarity, while recognizing the difficulties that emerge when reckoning with multidisciplinarity. As HERE has grown and changed, humanists have had to think like engineers, engineers have had to think like humanists, and we have all had to put ourselves in the mindset of first-year students at a STEM school. The multidisciplinary and interpersonal interactions experienced by HERE students, faculty, and staff have been among the most valuable and enjoyable opportunities afforded by the program. Although we have attempted interdisciplinarity, we have not reassessed our disciplinary assumptions, and if anything, our experiences in the HERE program have demonstrated the sturdiness of our disciplinary silos.

Teaching in and administering the HERE program therefore adheres to the scenario Louis Menand describes in *The Marketplace of Ideas: Reform and Resistance in the American University* (2010). Menand focuses on the humanities, but his observations about disciplinarity are applicable to our experiences. Menand charts the rise of humanistic departments in universities during the late nineteenth and early twentieth centuries. He remains skeptical of interdisciplinarity and its claims of rendering disciplines and academic departments unnecessary. He further doubts that transdisciplinarity would be preferable to the current configuration, if it can even exist. Anxieties about academic privilege and the academy's reluctance to change undergird interdisciplinarity, and despite its promise, it only accomplishes "the scholarly and pedagogical ratification of disciplinarity." Interdisciplinary arrangements result in "borrowed authority" instead of exchanges of knowledge between fields and the questioning of epistemological assumptions in the disciplines.[14] Interdisciplinarity in fact reaffirms and enhances disciplinarity.

In our attempts to be interdisciplinary, in other words, our identities as English professors at an engineering college have been brought into stark relief, to our students, our colleagues, and ourselves. Our teaching has not transcended the multidisciplinary: we remain English professors approaching a non-English discipline problem. Even when we try to convince our colleagues, in the program and elsewhere, that HERE and sustainability should be among the protagonists in the college's story, we have never felt more like English professors than when teaching HERE students, undercut by the narrative we hoped to change.

Sustainability Stories and the Entrepreneurial Ethos

No matter the type, all organizations tell stories that convey values and goals. These stories fit into larger narratives about an organization, and it is narrative that ultimately drives our passively unsustainable environment. Narrative can be understood as part of what Jennie Winter and Debby Cotton call "the hidden curriculum," which

> may consist of the values and beliefs of the institution or the individual lecturers which are unconsciously transmitted to the student, or impact on the institutional environment, thereby affecting student learning. In the case of sustainability, these impacts may be largely negative . . . mean[ing] that students receive incompatible messages about sustainability from different dimensions of their experiences.

At RHIT, innovation and entrepreneurship have recently become the college's reason for being. As we have said, these are not problems endemic to our college, to engineering education, or to other kinds of professional education. All universities sell themselves; they attract students by promising return on investment, a phrase too often defined economically. But in our institution, as in many others, faculty, staff, and students who pursue innovation and entrepreneurship enjoy numerous advantages over those who pursue sustainability. The institute, already a corporate-friendly environment, showcases a spacious innovation center for student design teams, a program (centered, like our failed Ouabache Project, on faculty development) for infusing entrepreneurship initiatives across the curriculum (funded by a $2.25 million grant from the Kern Family Foundation), a first-year living-learning community (modeled on the HERE Program, though from the top down, and funded by a $1 million grant from the Lilly Endowment) for students interested in developing an entrepreneurial mindset, and a new faculty line in entrepreneurship.[15]

Now, juxtapose the dedication to innovation and entrepreneurship—and the $3.25 million in private grant money—with the status of sustainability: The institute has no administrative champion for sustainability, but rather a Sustainability Team that never meets. It has no centralized academic support system for faculty members who are teaching or researching sustainability. HERE has a $20,000 yearly budget, half of which is external and must be reapplied for every year. The sustainability studies certificate program, which will likely be phased out by the end of the 2016–2017 academic year, graduated two students in May 2015. At the level of campus operations the picture improves, as the institute has a Director of Environmental Health and Safety, and a Department of Facilities Operations that has incorporated some aspects of sustainability into its daily practice, including an expanded on-campus recycling program. But the story suggested by this juxtaposition implies institutional indifference and missed opportunities.

As outlined above, sustainability and entrepreneurship have an uneasy relationship. Saving energy, making money, and helping others can coalesce—being entrepreneurial and sustainable are by no means mutually exclusive, as movements such as "natural capitalism" suggest. Yet with sustainability, the tensions between it and cultural narratives about engineering and the purposes of higher education often emerge, whereas with entrepreneurship the connections are less fraught. Many schools, including ours, stress business acumen nearly as much as technical expertise. In our experience, these strategies do help students consider matters like life-cycle costs, supply chains, and economic valuation of environmental services. They work less well for those areas of the Venn-diagrammed triple bottom line that define "value" in more than fiscal terms. Emphasis on disruptive innovation and entrepreneurship in some cases helps students who care about sustainability to bridge competing ideologies, doing well by doing good. However, in practice, it also often means the familiar reduction of the triple bottom line to a single, financial bottom line.[16]

The narrative we would substitute has the moral that the world's needs come first, and that we fail our students when we forget this. The language of sustainability is as susceptible to co-optation as is the language of corporate social responsibility since "discourses are, by their nature, problematic and contestable, open to interpretation and reinterpretation and governed by the motives of those who develop the discourse."[17] In an academic culture that too often stresses vocational and industrial traits over critical literacies and citizenship, entrepreneurship may just as easily distort the three pillars of sustainability as support them. The single bottom line of economic gain, even when submerged into the "value-added" rhetoric of entrepreneurship, fails to address social responsibility and environmental impact. Thus, tensions become visible between deep-seated cultural narratives, competing ideologies about the engineering profession, the purposes of higher education, and divergent definitions of sustainability. Entrepreneurship purports to be value neutral, whereas sustainability does not.

We therefore find ourselves in a double-bind. If we do not link sustainability to vocational prospects, students consider it irrelevant. When we reinforce the notion that the purpose of education is lucrative employment, that message is internalized by our students, and students' engagement with their education is unlikely ever to be personal or emotional. It is a mercenary relationship, utilitarian. When college exists only as vocational career preparation, college ought *not* to be transformative. Higher education's current master narrative, at RHIT and across the United States, leaves little room for community, and if we turn out students who care little for communities—their own and others'—then we cannot turn out students who truly understand sustainability. If the dominant ethos of higher education is fiscal gain, then a counter-narrative must arise. When the HERE program insists that students understand sustainability problems and their contexts, foster historical and cultural awareness, and consider the nontechnical elements of sustainability and higher education, we offer an alternative to fiscal narratives. The terms "innovation" and "entrepreneurship" could never be equivalent to sustainability; indeed, in our experience, innovation and entrepreneurship may render sustainability palatable, but they also leave it toothless.

Create and Follow Your Narrative

Sustainability does not yet exist across the curriculum at RHIT. The main obstacles to the full integration of sustainability stem from our failure to meet the expectations of first-year students, from the limits imposed by our disciplinary affiliations, and from the competing narratives of sustainability and monetary gain—the latter fueled by innovation and entrepreneurship. All three obstacles are distinct, yet what binds them is the disjunction between our desired and our contextual narratives. Incoming students see statistics about job placement and average starting salaries and twenty-year return on investment, and upon arriving join a sustainability program in which monetary gain is minimized. The institutional narrative does not square with the program narrative, hence student confusion or resentment.

HERE faculty bear some responsibility for our program's shortcomings. We soft-pedal education and sustainability by settling for "weak" sustainability—business as usual, just more frugal—and yet we lose students when we challenge them to consider "strong" sustainability, losing them because they misconstrue our challenges to their thinking as arrogant, and because strong sustainability threatens the higher-education narrative of the twenty-first century.[18] HERE nudges students to question the master narrative, and to reflect on whether social and environmental problems can be solved by innovation and entrepreneurship. Furthermore, hard-edged realities hamper our abilities to devote more time to HERE than we already do; the program is superadded to our teaching, advising, service, scholarship, and lives outside the institute's purview. Beyond our small group and a smattering of outstanding students, RHIT does not have academic interests in sustainability. How, then, can the reductive and all-consuming narrative of college as vocational training for profit be challenged?

Our answer may sound simplistic, but is in practice as complicated as our experience continues to show: by telling an equally persuasive story about sustainability and higher education in the twenty-first century. This story has to acknowledge the lucrative careers that await most engineers, while exposing average starting salary as a poor measure of success. In its place HERE must emphasize the values of environmental stewardship, social and ethical responsibility toward underserved populations, and the intrinsic worth of mitigating economic hardships and inequalities in the United States and throughout the world. Unlike the current narrative—which places profit so far above people and planet as to say in essence *people and the planet do not matter*—our counter-narrative puts profit below the exigencies of environmental and ethical concern. Such a story might help students see their work in more expansive contexts so that their understandable desire to make money is seen more properly as one tool among many for trying to meet the world's needs. And while it will make students more attractive to employers if they conceive of themselves and their work in multidimensional ways, their education is ultimately directed not toward individual profit but toward the common good.

Sustainability programs in situations like ours should offer compelling narratives about their roles—philosophical and practical—in themselves and in the larger contexts they reside in. HERE is just starting to tell the story of what it can do, and while a powerful narrative would not have eliminated all our problems, it might have mitigated them had our story been more clearly told in 2010. Our failure to develop a counter-narrative contributes to our unsustainable environment. At the same time, we worry that no counter-narrative may be able to compete with—or be heard alongside—the narrative of profit that is never subject to scrutiny.

Without a counter-narrative to challenge the profit narrative, undergraduate sustainability programs will find it difficult to attract and retain students, who are increasingly beset by materialist pressures. The need to make money, and more of it, is not going to vanish, but that does not mean that sustainability programs should be excluded from discussions of what constitutes value. The case must be made, strongly and often, that sustainability in any discipline deserves to be part of what higher education, and life beyond it, is for. By reclaiming higher education from the myopic narrative that equates money with success, sustainability initiatives will be more likely to succeed in hostile, or, as in our case, passively unsustainable environments.

NOTES

1. Our list of the impacts of HERE on campus is not an exhaustive list of sustainability-focused courses at RHIT or of sustainability initiatives on campus; instead, we here are highlighting things attributable to HERE's influence. See also Arthur Foulkes, "HERE, HERE! Students Leaving Sustainable Mark on Campus through Projects, Coursework

and Organizations," *Echoes* (Spring 2015): 18–19, http://www.rose-hulman.edu/media/1642469/Echoes_Spring2015.pdf.

2. "Rose-Hulman Taking Steps to Infuse Sustainability into Engineering Education," April 11, 2011, http://www.rose-hulman.edu/news/academics/announcing-a-new-fall-program-here.aspx; "HERE Member Is Finalist in International Sustainability Essay Contest," February 22, 2012, http://www.rose-hulman.edu/news/academics/2012/international-sustainability-essay-contest.aspx; Jane Santucci, "Your Green Valley: Rose Hopes to Engage Students to Live Sustainably," *Terre Haute Tribune-Star*, July 31, 2011, http://www.tribstar.com; Brian Boyce, "RHIT Students Plotting Ways to Help Environmentally," *Terre Haute Tribune-Star*, December 17, 2012, http://www.tribstar.com; Jennifer Aurandt et al., "Bringing Environmental Sustainability to Undergraduate Engineering Education: Experiences in an Inter-Disciplinary Course," *Journal of STEM Education* 13, no. 2 (April 2012): 15–24.
3. The pervasiveness of corporate rhetoric and management practices, and the growing number of higher-education administrators and executives with exclusively corporate backgrounds—and corollary deleterious effects on public and private universities—have been discussed widely. See, for example, Henry A. Giroux, "Neoliberalism, Corporate Culture, and the Promise of Higher Education: The University as a Democratic Public Sphere," *Harvard Educational Review* 72, no. 4 (Winter 2002): 425–463; and Jennifer Washburn, *University, Inc.: The Corporate Corruption of Higher Education* (New York: Basic Books, 2005).
4. Jonathan Rothwell and Siddharth Kulkarni, *Beyond College Rankings: A Value-Added Approach to Assessing Two- and Four-Year Schools* (Washington, DC: Brookings Institution, 2015), http://www.brookings.edu/research/reports2/2015/04/29-beyond-college-rankings-rothwell-kulkarni; "*The Economist* Lists Rose-Hulman 23rd Nationally, 1st among Indiana Colleges for Value," November 5, 2015, http://www.rose-hulman.edu/news/academics/2015/the-economist-lists-rose-hulman-23rd-nationally,-1st-among-indiana-colleges-for-value.aspx; "College ROI Report: Best Value Colleges," PayScale Inc., 2016, http://www.payscale.com/college-roi; "97% Placement Rate Affirms Return on Investment," Rose-Hulman Institute of Technology, January 9, 2017, http://www.rose-hulman.edu/news/on-campus/2017/97-placement-rate-affirms-return-on-investment.aspx.

   RHIT's scores in mid-career earnings, occupational earning power, and loan repayment rate are 100, 98, and 96, respectively, on a 100-point scale, according to the fully searchable Brookings report. However, Rothwell concedes that engineering colleges like Rose-Hulman score as highly as they do because of the general marketability of

STEM degrees: "STEM is the biggest measurable factor [and] it works even if you don't go to an elite school. . . . Even if you go to community college, you'll see an earnings premium." Rothman notes also that lifetime earnings are often predicted more effectively by test scores than by the college where a student matriculates. These observations also are cited in Ariel Schwartz, "Forget Harvard: Here's Where to Go to College if You Want a High Paying Job," *Fast Company*, April 30, 2015, http://www.fastcoexist.com/3045598/forget-harvard-heres-where-to-go-to-college-if-you-want-a-high-paying-job.

5. The ACUPCC was spearheaded by Second Nature and several corporate and university partners; see Second Nature Inc., http://secondnature.org/. Sustainability initiatives existed at RHIT prior to HERE, although they lacked (and continue to lack) unifying factors such as an office, an administrative champion, or a narrative. For an overview of previous efforts, see "Sustainability on Campus," *Echoes* (Summer 2008): 8–14, https://www.rose-hulman.edu/media/91125/echoes_summer_2008.pdf.
6. Two examples of successful cross-disciplinary faculty development workshops are Emory University's Piedmont Project and Northern Arizona University's Ponderosa Project. At RHIT, other administrators made similarly dismissive comments regarding later sustainability initiatives: One asked us why we would run HERE if it was not profitable, and another wondered if students would be interested in the sustainability studies certificate without a clear value proposition.
7. Matthias Barth, *Implementing Sustainability in Higher Education: Learning in an Age of Transformation* (London: Routledge, 2015), 117–118, 153.

   RHIT has reduced its carbon dioxide emissions from 22,213 metric tons in 2007 to 16,141 metric tons in 2015 (a reduction of 6,072 metric tons) and adopted a climate action plan in 2010—but neither of these facts have appeared in any campus news source, nor are they readily available on the RHIT website. Instead, this information is available via Second Nature, the organization responsible for the ACUPCC, at http://secondnature.org.
8. Donna Riley, *Engineering and Social Justice* (San Rafael, CA: Morgan & Claypool, 2008), 39; Matthew B. Crawford, *Shop Class as Soulcraft: An Inquiry into the Value of Work* (New York: Penguin Books, 2009).
9. David W. Orr, *Ecological Literacy: Education and the Transition to a Postmodern World.* (Albany: State University of New York Press, 1991), 24, 31.
10. Kevin Warburton, "Deep Learning and Education for Sustainability," *International Journal of Sustainability in Higher Education* 4, no. 1 (2003): 44–56, http://dx.doi.org/10.1108/14676370310455332. See also Katherine D. Arbuthnott, "Education for Sustainable Development beyond Attitude Change," *International Journal of Sustainability in Higher Education* 10, no. 2 (2009): 152–163, http://dx.doi.

org/10.1108/14676370910945954; Nicholas A. Ashford, "Major Challenges to Engineering Education for Sustainable Development: What Has to Change to Make It Creative, Effective, and Acceptable to the Established Disciplines?," *International Journal of Sustainability in Higher Education* 5, no. 3 (2004): 239–250, http://dx.doi.org/10.1108/14676370410546394.

11. Donald A. Schön, *The Reflective Practitioner: How Professionals Think in Action* (New York: Basic Books, 1983), 40.
12. Anne Steinemann, "Implementing Sustainable Development through Problem-Based Learning: Pedagogy and Practice," *Journal of Professional Issues in Engineering Education and Practice* 129, no. 4 (October 2003): 216–224, http://dx.doi.org/10.1061/(ASCE)1052-3928(2003)129:4(216).
13. See Mark Minster et al., "Sustainability and Professional Identity in Engineering Education," in *Higher Education for Sustainability: Cases, Challenges, and Opportunities from Across the Curriculum*, ed. Lucas F. Johnston (New York: Routledge, 2013); Richard B. Norgaard, "Transdisciplinary Shared Learning," in *Sustainability on Campus: Stories and Strategies for Change*, ed. Peggy F. Barlett and Geoffrey W. Chase (Cambridge, MA: MIT Press, 2004).
14. Louis Menand, *The Marketplace of Ideas: Reform and Resistance in the American University* (New York: W.W. Norton, 2010), 96–97, 119.
15. Jennie Winter and Debby Cotton, "Making the Hidden Curriculum Visible: Sustainability Literacy in Higher Education," *Environmental Education Research* 18, no. 6 (December 2012): 785–786, http://dx.doi.org/10.1080/13504622.2012.670207; "The Kern Family Foundation Awards $2.25M Grant to Rose-Hulman for Entrepreneurially-Minded Learning," Rose-Hulman Institute of Technology, November 12, 2014, http://www.rose-hulman.edu/news/academics/2014/the-kern-family-foundation-awards-$225m-grant-to-rose-hulman-for-entrepreneurially-minded-learning.aspx; "New Educational Initiatives Fostering an Entrepreneurial Mindset," Rose-Hulman Institute of Technology, September 25, 2014, http://www.rose-hulman.edu/news/academics/2014/new-educational-initiatives-fostering-an-entrepreneurial-mindset.aspx.
16. Richard House et al., "A Crooked Stool: The Rhetoric of the Triple Bottom Line," *Proceedings of the International Professional Communication Conference*, 2011, 1–4, ieeexplore.ieee.org/xpl/conhome.jsp?punumber=1000591.
17. Peter Dobers and Delyse Springett, "Corporate Social Responsibility: Discourse, Narratives, and Communication," *Corporate Social Responsibility and Environmental Management* 17, no. 2 (March–April 2010), http://dx.doi.org/10.1002/csr.231.
18. For a primer on weak and strong sustainability, see Jérôme Pelenc, Jérôme Ballet, and

Tom Dedeurwaerdere, "Weak Sustainability versus Strong Sustainability: Brief for Global Sustainable Development Report," *Sustainable Development Knowledge Platform,* United Nations, 2015, https://sustainabledevelopment.un.org/.

BIBLIOGRAPHY

Arbuthnott, Katherine D. "Education for Sustainable Development beyond Attitude Change." *International Journal of Sustainability in Higher Education* 10, no. 2 (2009): 152–163. Http://dx.doi.org/10.1108/14676370910945954.

Ashford, Nicholas A. "Major Challenges to Engineering Education for Sustainable Development: What Has to Change to Make It Creative, Effective, and Acceptable to the Established Disciplines?" *International Journal of Sustainability in Higher Education* 5, no. 3 (2004): 239–250. Http://dx.doi.org/10.1108/14676370410546394.

Aurandt, Jennifer, Terri Lynch-Caris, Andrew S. Borchers, Jacqueline El-Sayed, and Craig Hoff. "Bringing Environmental Sustainability to Undergraduate Engineering Education: Experiences in an Inter-Disciplinary Course." *Journal of STEM Education* 13, no. 2 (April 2012): 15–24.

Barth, Matthias. *Implementing Sustainability in Higher Education: Learning in an Age of Transformation.* London: Routledge, 2015.

Crawford, Matthew B. *Shop Class as Soulcraft: An Inquiry into the Value of Work.* New York: Penguin, 2009.

Dobers, Peter, and Delyse Springett. "Corporate Social Responsibility: Discourse, Narratives, and Communication." *Corporate Social Responsibility and Environmental Management* 17, no. 2 (March–April 2010): 63–69. Http://dx.doi.org/10.1002/csr.231.

Giroux, Henry A. "Neoliberalism, Corporate Culture, and the Promise of Higher Education: The University as a Democratic Public Sphere." *Harvard Educational Review* 72, no. 4 (Winter 2002): 425–463.

House, Richard, Corey Taylor, Jessica Livingston, Anneliese Watt, and Mark Minster. "A Crooked Stool: The Rhetoric of the Triple Bottom Line." *2011 IEEE International Professional Communication Conference (IPCC 2011)*. Red Hook, NY: Curran, 2012. Ieeexplore.ieee.org/xpl/conhome.jsp?punumber=1000591.

Menand, Louis. *The Marketplace of Ideas: Reform and Resistance in the American University.* New York: W.W. Norton, 2010.

Minster, Mark, Patricia D. Brackin, Rebecca DeVasher, Erik Z. Hayes, Richard House, and Corey Taylor. "Sustainability and Professional Identity in Engineering Education." In *Higher Education for Sustainability: Cases, Challenges, and Opportunities from Across the*

Curriculum, edited by Lucas F. Johnston, 109–123. New York: Routledge, 2013.

Norgaard, Richard B. “Transdisciplinary Shared Learning.” In *Sustainability on Campus: Stories and Strategies for Change*, edited by Peggy F. Barlett and Geoffrey W. Chase, 107–120. Cambridge, MA: MIT Press, 2004.

Orr, David W. *Ecological Literacy: Education and the Transition to a Postmodern World*. Albany: State University of New York Press, 1991.

Pelenc, Jérôme, Jérôme Ballet, and Tom Dedeurwaerdere. “Weak Sustainability versus Strong Sustainability: Brief for Global Sustainable Development Report.” *Sustainable Development Knowledge Platform*, United Nations, 2015. Https://sustainabledevelopment.un.org/.

Riley, Donna. *Engineering and Social Justice*. San Rafael, CA: Morgan & Claypool, 2008.

Schön, Donald A. *The Reflective Practitioner: How Professionals Think in Action*. New York: Basic, 1983.

Steinemann, Anne. “Implementing Sustainable Development through Problem-Based Learning: Pedagogy and Practice.” *Journal of Professional Issues in Engineering Education and Practice* 129, no. 4 (October 2003): 216–224. Http://dx.doi.org/10.1061/(ASCE)1052-3928(2003)129:4(216).

Warburton, Kevin. “Deep Learning and Education for Sustainability.” *International Journal of Sustainability in Higher Education* 4, no. 1 (2003): 44–56. Http://dx.doi.org/10.1108/14676370310455332.

Washburn, Jennifer. *University, Inc.: The Corporate Corruption of Higher Education*. New York: Basic Books, 2005.

Winter, Jennie, and Debby Cotton. “Making the Hidden Curriculum Visible: Sustainability Literacy in Higher Education.” *Environmental Education Research* 18, no. 6 (December 2012): 783–796. Http://dx.doi.org/10.1080/13504622.2012.670207.

Connecting Urban Students to Conservation through Recovery Plans for Endangered Species

Andrea Olive

Saving our civilization is not a spectator sport.

—Lester R. Brown, *Plan B 3.0: Mobilizing to Save Civilization*

Toronto was never a city I expected to live or work in. I was born on the prairies and spent my childhood visiting my grandparents' farm, skiing in the Rocky Mountains, and watching the sun rise across the huge open sky on my way to early morning swim practice. I was extraordinarily lucky to spend two whole months every summer on a small lake in rural Saskatchewan, in a cabin with a stove (and eventually a microwave), floor heaters for cold nights, and a telephone for emergencies. For almost sixty days a year, my best friends were the frogs and the loons. I learned to climb trees, build fires, outrun angry hornets, and swim in open water. The cabin housed us and Mom fed us, but nature sustained me.

After high school I moved to a big city for university. Living in Calgary was the first time I ever really experienced a true "traffic jam," and taking the Calgary C-Train to a hockey game was the first time I ever used public transportation. In Calgary, at the age of eighteen, I was finally able to see firsthand the impact of human beings taking over a large geographical space and creating a built environment. The effects on nature were not exactly at the forefront since Alberta is large in relation to the smallness of Calgary. But living in a sprawling city, I could feel the stress of urban life, and I witnessed a society's innovative responses: C-Train expansion, increased urban green space, a pedestrian-only downtown core, a running path along the Bow River, and cheap bus tours of Banff National Park on weekends. Despite these efforts, I nonetheless felt that Calgary was not a sustainable city—or at least it did not sustain me for long as I moved immediately upon graduation.

Halifax, Nova Scotia, was my home for only a year, and its sprawl grew by the day. The city is close to agricultural fields, fruit orchards, and the ocean. I shopped at a farmers' market for the first time in my life (as stealing peas from my grandparents' farm does not exactly constitute "shopping"). And while Halifax is a walkable city, I had no desire to see more buildings or explore an "urban park." My car was my escape, and I would traverse the beautiful island and Atlantic Ocean shoreline every weekend. After earning a master's degree, I spent the next year in Washington, DC, and felt the crushing weight of a "commute" five days a week. I had to catch a bus in Alexandria, Virginia, around seven so I could catch the Metro at Pentagon Station and travel downtown, where I would walk the last fifteen minutes to Dupont Circle. I spent about three hours a day commuting to my place of work. That lasted seven months before I left my job and got on a plane headed home to the prairie sunset.

Purdue University, where I earned my PhD, is in a small town in Indiana. The cornfields reminded me of home, and I settled in quickly. A year into my program I switched from studying political philosophy to environmental policy. The reason for a shift in focus was because a very small—one ounce—and very endangered bat was in need of help. The Indiana brown bat (*Myotis sodalis*) had been pushed aside for the expansion of the airport in Indianapolis, and it now required local landowners' stewardship on private land to survive. I wanted to find out how the government could help foster such stewardship on private lands to conserve the public good of biodiversity. I carried this project with me to Detroit, where I worked as an assistant professor at the University of Michigan, Dearborn. For all its problems, Detroit is a city in the midst of rebuilding with urban sustainability at the forefront. Urban farming, public transportation, and urban reclamation are central to a new Detroit. While there, I studied landowner attitudes toward urban trees, and continued to focus on the puzzle of endangered species conservation. Biodiversity is put at risk through cities—habitat destruction either through the expansion of a city's footprint or through the intensive agriculture necessary to feed it causes the crisis of endangered species. Yet those species depend on rural populations and rural land to survive. It is rural landowners who must take up the responsibility of stewarding the nature that was pushed out of cities.

The Greater Toronto Area, or GTA, is where I now live and work. It is by far Canada's largest city, and I can feel the vastness every time I get in my car. The 401 Highway that stretches across the northern metropolitan area is one of the busiest highways in the world. Sometimes I wonder if there are more people stuck in traffic on that highway than there are people in my home province of Saskatchewan. I teach on two campuses at the University of Toronto, Mississauga, and a three-hour commute is once again part of my life. (Thankfully it is not a daily commute.) This urban environment is the right place continually to find the inspiration to study endangered species conservation and sustainable living. And it is exactly the right place for me to stress the importance of these issues to my students. With these considerations and contexts in mind, this chapter provides an overview of my senior seminar course at UTM, titled "Conservation Policy in Canada." This course is important to me both professionally and personally, as I hope that my class can contribute to the protection and recovery of endangered species in the rural places of this country that I hold dear to my heart.

Biological and Political Assessment

When people think about Canada they often conjure up images of forests, lakes, mountains, and wide-open spaces. This imagined terrain is filled with moose, beavers, birds, and bears. In many respects this postcard picture is accurate. But it fails to paint the whole picture. Canada is the second largest country in the world in landmass and has fewer residents than the state of California, yet the country is predominantly urbanized. About 80 percent of Canadians live in a city. And those cities, like cities everywhere, cause environmental degradation. While Canada is home to an estimated seventy thousand species, biodiversity is still put at risk across the country through urbanization and agriculture. Indeed, urban development (including urban sprawl) is one of the largest threats to biodiversity in developed countries.[1] Right now Canada has more than 650 species at risk of extinction or expatriation, and this number is increasing with time.

In 1992, Canada signed the United Nations Convention on Biological Diversity and committed to the conservation and sustainable use of biological diversity and the fair and equitable sharing of the benefits arising from the use of genetic resources.[2] However, since then, the political reality of Canada's federal system has constrained progress. The country has a national government in Ottawa as well as ten provincial governments and three territorial governments. The Canadian Constitution grants different powers to the levels of government. With regard to wildlife, Section 91 of the Constitution grants the national government jurisdiction over international treaty making, federal land, migratory birds, aquatic species, and the oceans. Much more is left to the provinces, which are responsible for provincial lands and the regulation of private property. Moreover, the Canadian federal government is not a significant land manager across the ten provinces. In fact, only about 5 percent of land outside the three Northern territories is federal land—leaving 95 percent of the land to provincial governments to oversee, manage, and otherwise regulate.[3] Ontario, for example, is home to 186 federally listed endangered species. Like other provinces, Ontario has its own Endangered Species Act and maintains its own list of species. On its current list are just over two hundred provincially listed endangered species.

Recognizing the need for cooperation in the area of biological conservation, especially for at-risk species, the federal government convened a national meeting in 1996. The Accord for the Protection of Species at Risk evolved from the meeting, and all territorial and provincial governments (save Quebec) agreed to cooperate with the federal government in the protection of at-risk species.[4] After a six-year struggle in Parliament, the federal government passed the Species at Risk Act in 2002. This law does protect endangered species across the country, but it applies only to migratory birds, aquatic species, and wildlife found on federal land. Thus it is fairly limited, because of the constrained constitutional authority granted to the federal government. Nevertheless, the national government does maintain a list of species at risk and is responsible for writing recovery plans and action strategies for each endangered and threatened species—even those not found on federal lands. To date, more than six hundred species are on the list, a number that would likely be higher if the federal government moved more efficiently through the listing process, and if the independent agency responsible for wildlife assessment, called the Committee on the Status of Endangered Wildlife in Canada (COSEWIC), had the resources to assess more quickly Canada's flora and fauna.

Of the species assessed and listed in Canada, many do not have a recovery plan and/or action plan in place. Under the law, the government's recovery plan for each species must outline a scientific assessment of the species and make recovery recommendations on the basis of the best available knowledge, including science, local knowledge, and Aboriginal traditional knowledge. After the recovery plan is approved, the government must write an action plan for each species, in which policy recommendations consider economic, social, and political factors. As of summer 2015, the government has written and finalized 198 recovery strategies but only thirteen action plans. This shortcoming is in part because of the political, economic, and scientific complexity of recovering an endangered species, and in part because former prime minister Stephen Harper's Conservative Party–led federal government from 2006 to 2015 did not show much interest in (or dedicate the necessary resources to) fully implementing the Species at Risk Act. In response to pressure from environmental agencies and recent litigation, Harper's government announced a three-year plan to write a recovery strategy for every federally listed species. Overall, though, the process is slow, and the reality is that very few listed species in Canada have a recovery strategy and action plan in place. This lack of actual government action was the inspiration for me to create an assignment that asks my students to write these action plans for species that have already been recognized as imperiled enough to require a recovery strategy.[5]

Critical Habitat

The University of Toronto is the largest university in Canada and one of the largest in the world, with 68,000 undergraduate students and 16,000 graduate students. To accommodate the sheer number of students and the classroom space they require, the university has three campuses: one in the heart of downtown Toronto (St. George), one in the eastern part of the city (Scarborough), and one in the western suburb of Mississauga. Students and faculty regularly move among the three campuses, often by campus shuttle buses or personal vehicles. The distance between the Mississauga campus and St. George is thirty-three kilometers, and the distance between the Scarborough campus and St. George is thirty kilometers; during regular weekday traffic, those commutes can vary between forty-five and ninety minutes. The three individual campuses each have sustainability initiatives, and in 2016 the university was named one of the "greenest employers" in Canada. Yet transportation between the three campuses continues to put a major strain on the environment as it creates pollution and contributes to sprawl. What's more, the campus fails to address the issue of biodiversity loss and endangered species, with no mention whatsoever of these issues in the University Sustainability Yearbook, for example.[6]

The University of Toronto is not unique in this aspect. When it comes to biodiversity loss, universities, the majority of which exist in an urban context, are part of both the problem and solution. College campuses contribute to habitat loss as they take over large chunks of land, and colleges are responsible for drawing thousands (or more) of people into a city. However, institutions of higher education also, ironically, produce inside their departments of philosophy, history, geography, economics, political science, biology, and other related fields the research necessary to address biodiversity loss. This tension between the physical space of a campus and the research and learning inside that space makes universities a promising place to teach individuals about the connection between their lifestyle and its environmental impacts.

The university's home city, Toronto, is the capital of Canada's most populated province, Ontario. The city and its surrounding metropolitan area is the largest in Canada and the fourth largest in North America. The city itself has 2.6 million people, while the Greater Toronto Area (GTA) has six million people, representing about 15 percent of Canada's total population. The city of Toronto borders Lake Ontario and is intersected by three rivers. Canada's largest urban park, Rouge Urban Park, resides in the eastern part of the city (close to the university's Scarborough campus), and densely forested ravines create small parks throughout the GTA. Beyond that, there are numerous large parks in the downtown area as well as on the islands just off the downtown core's harbor front. To protect agricultural land and wildlife areas as well as prevent continuing urban sprawl, the city is buffered by a "greenbelt," a provincially protected green space created in 2005.[7] While the greenbelt is my weekend reprieve, I do spend my days caught in traffic on one of the world's busiest highways (the 401) or hopping between the GO Train and the metro system. It is an almost ninety-minute commute to my downtown office or a twenty-minute drive to my UTM office.

With so many people and so much traffic, it is easy to overlook the fact that Ontario is very rich in biodiversity. Toronto is found in the Mixedwood Plains ecozone in the northernmost part of North America's Carolinian zone, a deciduous forest region that extends to the Carolinas in the United States. This part of Canada is home to a lot of flora and fauna, despite increasing urbanization and agriculture in the area. According to the Ontario Ministry of Natural Resources and Forestry, the agency responsible for enforcing the province's Endangered Species Act, there are twenty-one species listed in Mississauga and eighteen species listed in Toronto (which includes Scarborough) on the provincial endangered species list—meaning the university shares its space with numerous endangered species.

Unfortunately, urban residents in Canada are not familiar with endangered species. Through surveys of Toronto and Vancouver residents, I discovered that only 24 percent of respondents had heard of the Species at Risk Act, and that only 37 percent could name an endangered species in the country.[8] Not surprisingly, students in my conservation policy course likewise are unfamiliar with conservation issues. At the start of the course, only students who took my second-year environmental policy course report knowing about the Species at Risk Act, and many students struggle to name an endangered species in Canada. The reasons for endangered listing—mainly habitat loss and climate change—are not salient among students. Indeed, they struggle to link their urban lifestyles to biodiversity loss in Canada. Most students at the University of Toronto campus in Mississauga grew up in the GTA, and most have rarely traveled outside urban areas in Canada. This is all to say that prior to enrolling in the course, students have given little thought to biodiversity loss, and they are virtually unaware of existing policy to recover at-risk species. It is for these reasons that I focus the course on the recovery-plan assignment and attempt to connect urban students to their responsibility of stewarding a public good like biodiversity.

The Recovery Plan

The purpose of my course is to teach students about conservation policy in Canada, primarily through reading and discussing strategies for endangered species conservation, including regulation, economic incentives, voluntary action, and education programs. The main assignment in the course, which is taught at the 400 level in a small seminar setting, is for each student to write a recovery strategy for a federally listed endangered species that does not presently have a plan in place by Environment and Climate Change Canada. In order to promote variation in taxonomy and geography, I choose the species and assign one to each student. This means there is at least one bird, mammal, amphibian, reptile, moss, tree, plant, and insect assigned in the course. Species range also varies across the country such that all ten provinces and three territories are part of the course. Since most students grew up in the GTA it is important for them to consider how other regions in Canada think about and rely on biodiversity. For example, in some places in Canada, First Nations play a vital role in land management, and their knowledge and participation is required for successful conservation. In other places, local communities are dependent upon fisheries or forestry for economic development, and approach endangered species conservation with a different set of values than urban residents. Exposing students to Canada at large helps them learn not just about the country but also about different values and the need for a variety of policy tactics to foster stewardship.

The recovery strategy assignment is a scaffolded assignment in which students slowly build a plan over the course of the twelve-week semester. The assignment is completed in seven parts: (1) summary of scientific assessment, (2) description of species and its needs, (3) threat identification, (4) recovery recommendations (science), (5) policy approaches for recovery, (6) presentation, and (7) final recovery strategy. Essentially, the project begins with the scientific literature around the species and its habitat, and it then builds to a set of policy recommendations for recovering the species in Canada. Since I developed this course in 2012 and taught it in winter 2013 and 2015, twenty-five students have written recovery plans for twenty-five different species, none of which had a federal plan published by the government of Canada. (See table 1 for species name, Latin name, taxonomy, federal status, and geographical range.)

The process of summarizing the existing scientific data for the species can be quite overwhelming for students, but it is often because so little information exists, as opposed to too much. Many students, especially those working on plants and insects, find that there is virtually nothing known about the species. All students begin by examining the official report by the Committee on the Status of Endangered Wildlife in Canada, but even this agency often has little information about a species, because less than 10 percent of Canada's biodiversity has been assessed by scientists. Students then must delve into scientific journals and try to find any and all research concerning their species. For species like the White Shark or the Peary Caribou, students find a quantity of independent, peer-reviewed scientific information to wade through. But for other species, such as the Blue-grey Taildropper or the Seaside Bone Lichen, extremely little is known about the species, its needs, or what threatens it. One student was so shocked and disheartened about the lack of information on her assigned species, the Bert's Predaceous Diving Beetle, that she used the information she collected for her assignment as a basis for a *Wikipedia* entry, labor she undertook completely voluntarily and outside the course. She

TABLE 1. Canadian Species at Risk Assigned to Students in Course

COMMON NAME	LATIN NAME	TAXON	FEDERAL STATUS	RANGE
Cherry birch	*Betula lenta*	Vascular plant	Endangered	ON
Western rattlesnake	*Crotalus oreganus*	Reptile	Threatened	BC
Bicknell's thrush	*Catharus bicknelli*	Bird	Threatened	NB, NS, and QU
Bert's Predaceous Diving Beetle	*Sanfilippodytes bertae*	Arthropod	Endangered	AB
Rusty-patched Bumble Bee	*Bombus affinis*	Arthropod	Endangered	ON, QU
Verna's Flower Moth	*Schinia verna*	Arthropod	Threatened	AB, MN, and SK
Seaside Bone Lichen	*Hypogymnia heterophylla*	Lichen	Threatened	BC
Blue-grey Taildropper	*Prophysaon coeruleum*	Mollusk	Endangered	BC
Black-tailed Prairie Dog	*Cynomys ludovicianus*	Mammal	Threatened	SK
Rocky Mountain Tailed Frog	*Ascaphus montanus*	Amphibian	Threatened	BC
Great Basin Spadefoot	*Spea intermontana*	Amphibian	Threatened	BC
Porsild's Bryum	*Haplodontium macrocarpum*	Mosses	Threatened	AB, BC, NL, and NU
Tri-colored Bat	*Perimyotis subflavus*	Mammal	Endangered	NB, NS, ON, and QU
Wood Bison	*Bison bison athabascae*	Mammal	Special Concern	AB, BC, MN, NT, and YT
Blue Racer	*Coluber constrictor foxii*	Reptile	Endangered	ON
Peary Caribou	*Rangifer tarandus pearyi*	Mammal	Endangered	NT and NU
Golden-winged Warbler	*Vermivora chrysoptera*	Bird	Threatened	MN, ON, and QU
Desert Night snake	*Hypsiglena chlorophaea*	Reptile	Endangered	BC
Gold-edged Gem	*Schinia avemensis*	Arthropod	Endangered	AB, MN, and SK
White Shark	*Carcharodon carcharias*	Fish	Endangered	Atlantic Ocean
Greater Short-horned Lizard	*Phrynosoma hernandesi*	Reptile	Endangered	AB and SK
Western Screech Owl	*Megascops kennicottii kennicottii*	Bird	Threatened	BC
Western Painted Turtle	*Chrysemys picta bellii*	Reptile	Endangered	BC
Whitebark Pine	*Pinus albicaulis*	Vascular plant	Endangered	AB and BC

simply felt compelled to document the existence and known body of scientific literature for easy reference on the public database. Now when someone Googles this species, her entry comes up and people can more easily learn about the beetle.[9]

Using only scientific information, the students must describe their species and its habitat needs. For this stage of the assignment, pictures, maps, and other visual aids are important. It is often at this stage that students become increasingly "attached" to their species. Most students initially care little for an "ugly" frog, tiny insect, or a tree, disappointed that their assigned species is not more charismatic. But once students learn how unique their species is and the role it plays in a larger ecosystem, they become advocates for its protection. By semester's end, each student claims that his or her assigned species should be the "flagship species" of the class. Teaching students about the value of all biodiversity—charismatic or not—is a key component of the course.

At the next stage of the assignment, the threat identification stage, it is important that students use only the scientific literature and create an exhaustive list of both the direct and indirect threats to the species' survival. (See table 2 for an example of a student's threat summation for the Blue Racer snake on Pelee Island in Ontario.) Thus far, every species assessed in the class has "habitat loss or fragmentation" identified as a threat. Urbanization is not the only cause, as forestry, mining, oil extraction, or other resource extraction threatens habitat for many species. Nevertheless, the connection between human activity and species rate of decline is apparent, making habitat protection an issue the students must confront in this assignment.

Following the threat identification stage, students must also create a list of corresponding recommendations to address each threat. Here again, the objective is to use only scientific data and not to introduce economic or political dimensions. All recovery recommendations at this stage are scientific, but at the next stage students must jump to political science. To address the threats facing the Gold-edged Gem, a small moth native to Canada's prairie provinces, one student compiled this list of recommendations:

1. Develop an improved understanding of the demographics and reproductive cycles.
2. Identify and implement protocols that mitigate factors contributing to the steady decline of the active sand dune habitats.

TABLE 2. Threats Facing the Blue Racer Snake, 2013

THREAT CATEGORY	EXTENT OF THE THREAT	LEVEL OF SEVERITY
Habitat loss and fragmentation	Habitat conversion and fragmentation and loss of suitable habitat. Reduced resource availability and ability to migrate and/or return to den.	High
Road mortality	Accidental and intentional mortality from being struck by motor vehicles on roadways leading to decreased population size.	Medium
Vegetation succession	Controlled and uncontrolled burns on habitat leading to decreased cover and increased prey mortality.	Low
Small population size	Isolation from other populations due to fragmentation of habitat.	Low
Wild Turkey threat	The release of a predator species.	Low
Human disturbance of individuals	Disturbance or harm. Recreational visits and tourist activities as well as industrial activities. Intentional kills. Behavioral changes, reduced ability to migrate and/or return to den, and reduced population size.	Low

3. Identify, maintain, enhance, and increase habitat, along with maintaining and increasing the species that the Gold-edged Gem depends on in the habitat, namely, prairie sunflowers and skeleton weed.
4. Optimize the survival of larvae on the host plant (Prairie sunflower) through all seasons.
5. Identify causes of successive or invasive vegetation in the historical habitats and prevent them from occurring in the current habitat.
6. Identify and evaluate similar habitats that could be converted to recovery habitats in order to bridge the gaps between disjointed or isolated habitats.
7. Preserve and protect current, historical, and future recovery areas.
8. Encourage management, conservation, and research of the Gold-edged Gem and the habitats it uses in the United States.
9. Engage, support, and communicate with landholders and land managers about actions that may improve Gold-edged Gem populations and habitats in their local areas.
10. Promote and increase awareness about the Gold-edged Gem population to the general public.

TABLE 3. Student's Policy Recommendations to Recover the Desert Night Snake, 2015

POLICY APPROACH	OBJECTIVES
Field research	• Construct comprehensive habitat map • Delineate area of critical habitat • Characterize land usages within range • Collect landowner contact information
"Canada's Rarest Snake" education campaign	• Develop ecoliteracy • Foster civic engagement • Enhance volunteer capacity • Exert pressure on BC's provincial government • Eliminate threat of human persecution
Voluntary Stewardship Program	• Maintain critical habitat without use of financial incentives • Inform landowners of extirpation threat
Consult BC's provincial government with recourse to litigation	• Intervene to ensure ecological justice • Persuade government to acquiesce in remaining policy approaches
Aim to eliminate further agricultural and residential expansion into critical habitat	• Put a stop to further habitat fragmentation • Preemptively circumvent the need to compensate landowners for foregone usage
Seek to remove sourcing of aggregate construction materials throughout range	• Avert further habitat degradation due to local construction
Install signs along BC highways and roadways throughout range	• Reduce risk of accidental automobile mortality

In order to make policy recommendations, students must consider elements of financial cost and sociopolitical aspects of recovery, such as who owns the land. Indeed, a large portion of the course is devoted to endangered species recovery on private lands because these are where the majority of species are found in Canada. Students also must consider whether their recommendations are politically and economically feasible by consulting relevant literature. (See table 3 for an example of a student's policy recommendations to recover the Desert Night snake, a very rare snake found in Southern British Columbia.)

The final two stages of the assignment ask students to pull together each previous step and make a ten-minute class presentation about their species. After receiving feedback from the class and me, students take two weeks to compile their project and hand in a summative paper that acts as both a recovery strategy (the scientific part) and an action plan (the policy part). These plans range from ten to thirty pages depending on the amount and type of information known about the species.

Connecting Plans and People

After the students complete the final draft of their species' recovery plan and submit them to me in a PDF document, I email the plans to relevant scientists as well as Environment and Climate Change Canada and appropriate provincial ministries of Environment or Natural Resources. I also send the recovery strategies to Ecojustice, the David Suzuki Foundation, and other organizations or people relevant to the species. The response has been encouraging. While not all recovery strategies receive an official response, two examples are worth describing.

First, the Blue Racer is an endangered snake in Ontario that is similar in context to an endangered snake I studied (the Lake Erie water snake) in my own research. Having numerous contacts in this area, I emailed the student's strategy to a colleague at Ecojustice, a colleague at the David Suzuki Foundation, and a few scientists at the Ontario Ministry of Natural Resources and Forestry (MNRF). The director of Ecojustice emailed me back within a few hours to say, "I've circulated to our Species at Risk focused science staff. Great project idea." Another Ecojustice representative also emailed to say, "It is GREAT that your students do recovery strategies. I am totally keen to read them." At a more direct policy level, a species-at-risk biologist from MNRF replied that the "study may be relevant to the development of the recovery strategy and associated government response statement under the ESA." Finally, when I visited the MNRF in the summer of that year to discuss my own research in the province, a biologist to whom I was introduced recalled the Blue Racer recovery strategy that my student wrote and commented on its quality and usefulness.

In the case of the Desert Night snake, I emailed the document to Environment Canada, two members of the recovery strategy team, and the Canadian Wildlife Service. Within a few days, I received a number of responses. As one example, a recovery team specialist replied with enthusiasm, saying that the student

> did a great job given it's his first (presumably) attempt at developing a Recovery Strategy. The information uncovered is quite inclusive. . . . While not all of the [student's] Recovery Recommendations are feasible (we are limited in recovery

> efforts for any number of reasons), some are similar to our own recommendations and some are interesting suggestions that may have some merit.

The email went on to ask permission to "borrow some elements" from the student's plan. As a result of these email exchanges, the British Columbia Ministry of Environment, in collaboration with Environment and Climate Change Canada, has established an informal partnership with the course to target species in need in the BC/Yukon region. Thus, the next time I teach the course (winter 2017), the Ministry will choose a few species with which it needs assistance and ask my students to write their recovery assignments for those species.

Recovery Recommendations

Urban students need to be intellectually connected to biodiversity loss and the impacts of urbanization so that risks and environmental costs can be minimized. Toward the end of the course, the class discusses the role of awareness and education in species-at-risk protection and recovery. We at look at the work of nongovernmental organizations and discuss topics like smart growth, restoration ecology, and rewilding. By this point, students have realized the connection between species habitat and species survival. There is only so much space in Canada and on the planet; where University of Toronto students live—in the northern part of the deciduous forest close to a Great Lake—is the exact same place where a variety of other species live. Across the planet, human beings and other species are competing for the same space.

Every species in Canada deserves a recovery plan, yet the Canadian government has been slow to develop these plans. My course exposes this gap and pushes to fill the void. By writing a recovery plan for a species, students are immediately drawn into the world of conservation policy where they must face science, politics, and economics. The protection and recovery of at-risk species is a challenge, and students are asked to come up with creative ways to address the forces that threaten their species' survival.

This course can easily be modified to work across different scholarly fields and in different countries. Presently, the course is part of the geography department curriculum and is taught by a political scientist. However, the course could undoubtedly be co-taught by a geographer and a biologist, for example, or a political scientist and ecologist. Such pairings would offer more depth on the scientific side and link the science and policy more directly (especially as students move from a scientific recovery strategy to a policy-oriented action plan).

The course could also be aimed at provincial or territorial recovery plans as opposed to federal plans. For example, students in my course could focus solely on Ontario species at risk. This approach could work in virtually any province, except the handful that do not have stand-alone provincial legislation and are not presently developing provincial recovery documents (such as Saskatchewan or Prince Edward Island). That said, perhaps it is more important than ever to work inside the provinces that are struggling (with political will and resources) to list species, write plans, and implement strategies to recover species. British Columbia's government, for one, already has asked the class for some assistance. It is likely that provinces could also use some encouragement and assistance from competent students.[10]

Last, this course also can be easily adapted to other countries where endangered species exist. In the context of the United States, the Fish & Wildlife Service is responsible for listing species and writing recovery strategies, in a single document for each species. While the agency has been efficient at producing recovery strategies, many now are outdated and in need of revision. Species that were listed in the 1970s after the Endangered Species Act became law, for instance, might have recovery strategies that are dated 1973 or 1974—an inadequate reality for species facing increasingly difficult threats. An excellent student project would be to update the scientific information in these strategies and make new policy recommendations based on the forty years' worth of endangered-species protection and recovery experience in the United States.

Conclusion

Human beings are finding it increasingly difficult to share habitat and coexist with other living things on the planet. This reality is nowhere more apparent than inside cities. The rapid process of urbanization—a global phenomenon—is creating larger and larger cities while destroying and fragmenting the natural landscape. Such new "human dominated landscapes" make very little space for biodiversity to thrive, as Dale D. Goble, J. Michael Scott, and Frank W. Davis demonstrate. The result has been a steady rise in endangered flora and fauna and the mass extinction of species across the globe, what Richard Leaky and Roger Lewin call the "sixth extinction."[11] Educating students about biodiversity loss is a critical component of reversing the downward trend. The world needs individuals who understand the importance of species habitat and are able to find innovative ways to protect habitat in otherwise unsustainable human-dominated landscapes.

Every endangered species deserves a recovery strategy and action plan. So long as the federal and provincial governments are slow to write and implement plans in Canada, my students will continue to address the gap. While the plans are certainly not "implementation-ready," the students' work is a step in the right direction—and a reminder to the government and other organizations that these species are still waiting and that citizens do care about this issue. Even though I have traded my prairie sunsets for an urban commute, I care about this issue. And I will continue to teach urban students about sustainability and biodiversity loss inside the city and beyond the fringe.

In fact, teaching environmental and conservation policy at the University of Toronto is how I am able to survive in a region of six million people. I am able to share my research with hundreds of students each year in hopes that it will make a difference. Thus far the response from government agencies, scientists, and environmental groups suggests that I am on the right track. University of Toronto Mississauga students are contributing to a more sustainable Canada. This is critical for me as a scholar and as a Canadian. In regard to the latter, and perhaps selfishly, I need urban Canadians to care about prairie birds because I still spend my summers at the same lake I did as I child. I live like a migratory bird. I am a prairie grassland obligate and I have to return home every year in order to survive. Toronto now houses me and the Ontario Greenbelt now feeds me, but it is that small lake on a prairie field that has always sustained me. I hope all Canadians have a piece of nature they want to protect, and I believe that working together we can find new and innovative ways to reverse unsustainable trends and protect the places that have always sustained us.

Appendix: Recovery Strategy Assignment

Over the course of this semester you will be required to write a recovery strategy for your endangered species. This project will be completed in SEVEN steps. Each step will be handed in separately (step 6 will be a presentation to the class). Below is a list of the steps as well as directions for each component.

1. Summary of COSEWIC Assessment
2. Description of Species and Its Needs
3. Threat Identification
4. Recovery Recommendations
5. Policy Approaches for Recovery
6. Presentation
7. Final Recovery Strategy

In order to complete these assignments you will need to look at a few examples of written recovery strategies available on the Ministry of Environment and Climate Change Canada website.

1. SUMMARY OF COSEWIC ASSESSMENT. In one page (typed single-spaced), provide a summary of COSEWIC's most recent assessment of your species. This must include the date of assessment, the common name, scientific name, COSEWIC status, reason for designation, Canadian occurrence, and assessment history.

2. DESCRIPTION OF SPECIES AND ITS NEEDS. In three to five pages (typed single-spaced), provide a description of your species as well as its needs for survival. You will need to include a map (and written description) of your species' habitat. Please be sure to include a works cited (which is not part of the overall page total). You will need to consult numerous sources for this assignment, including articles in academic journals (use Web of Science database). You may include a picture of your species, but keep the picture small (no more than half a page).

3. THREAT IDENTIFICATION AND DESCRIPTION OF CRITICAL HABITAT. In two to five pages (typed single-spaced), provide a description of the known threats to your species' survival. It is important to be comprehensive and clear because later your recovery recommendations and policy approaches will have to address the mitigation of these threats.

You must also outline what you think would be critical habitat for this species. Obviously this will not be exact since you are not going to go out and survey the land. But based on the information you have about your species, what might the critical habitat entail? Include a map if necessary. Be sure to include a works cited.

4. RECOVERY RECOMMENDATIONS. In three to five pages (typed single-spaced), describe what is necessary for the recovery of your species. In this assignment pay no attention to the cost or political feasibility of your recommendations. Instead, use science as the guiding logic to address species protection and recovery. Be sure to include a works cited.

5. POLICY APPROACHES FOR RECOVERY. In three to six pages (typed single-spaced), lay out the policy tools that you would recommend to achieve recovery of your species. Here you must think like a politician and consider the costs and real-world feasibility of recovery. Include as many policy tools from class as necessary.

6. PRESENTATION. During class each student will be given 10 minutes (and not a second more) to present their recovery strategy. You may use any format you want (handouts, PowerPoint, Prezi, etc.). You must quickly summarize assignments 1 through 3 and then explain step 5 in more detail.

7. FINAL RECOVERY PLAN. Pulling together and refining steps 1 through 5, you must write a 10–15 page (typed single-spaced) recovery strategy for your species. The final paper must be polished. You should include pictures and maps. Also, be sure to include a works cited. Look at examples of action plans as well as recovery strategies to help create this final piece. Ideally, I would like to send your completed recovery strategy to the Recovery Team leads at Environment Canada.

NOTES

1. Stephen DeStefano and Richard M. Degraaf, "Exploring the Ecology of Suburban Wildlife," *Frontiers in Ecology and the Environment* 1, no. 2 (March 2003): 95–101. See also Office of the Auditor General, *Report of the Commissioner of the Environment and Sustainable Development*, chapter 1, "Backgrounder on Biological Diversity" (Ottawa: Office of the Auditor General, 2013); and John M. Marzluff, "Fringe Conservation: A Call to Action," *Conservation Biology* 16, no. 5 (October 2002): 1175–1176.
2. United Nations, Environment Programme, *Convention on Biological Diversity* (1993). See also Environment Canada, "Welcome to Biodivcanada.ca," http://biodivcanada.ca.
3. The relationship between Aboriginal peoples and conservation of species at risk in Canada is very complicated. There are about 650 First Nation bands in Canada that hold 28,000 square kilometers of reserve land. But that land technically is part of the "Federal House" and falls under federal jurisdiction. Thus, the Species at Risk Act applies to First Nation reserves. Accordingly, the law carves out a large role for First Nations to play in the

listing, recovery, and protection of species at risk. In the Northern territories, most of the land is also federal, but recent land-claim agreements between the Inuit and the federal government provide some Inuit involvement in conservation on Inuit lands. The creation of wildlife boards that comanage wildlife with the territories, the federal government, and the Inuit has been a result of these land-claim agreements. See Andrea Olive, *Land, Stewardship, and Legitimacy: Endangered Species Policy in Canada and the United States* (Toronto: University of Toronto Press, 2014).

4. The territory of Nunavut was carved out of the Northwest Territories and became its own territory in 1999. Because it did not exist in 1996, Nunavut thus is not technically a signatory to the accord.
5. Species at Risk Public Registry Office, "Recovery Strategies," Government of Canada, Environment Canada, http://www.registrelep-sararegistry.gc.ca/sar/recovery/recovery_e.cfm; Olive, *Land, Stewardship, and Legitimacy*, 49–73. Only time will tell if the 2015 election of Prime Minister Justin Trudeau and the Liberal Party will mean a different approach to implementing the Species at Risk Act and writing recovery strategies/action plans.
6. University of Toronto, "Quick Facts," http://www.utoronto.ca/about-uoft/quickfacts; Richard Yerema and Kristina Leung, "University of Toronto Recognized as One of Canada's Greenest Employers (2016)," Mediacorp Canada Inc., http://content.eluta.ca/top-employer-university-of-toronto; University of Toronto, Facilities and Services, Sustainability Office, *2015–2016 Sustainability Yearbook: Highlights from St. George Campus*, http://www.fs.utoronto.ca/sustainability-office/sustainability-yearbook-2015-16/.
7. Statistics Canada, "Population of Census Metropolitan Areas," Government of Canada, http://www.statcan.gc.ca; Ministry of Municipal Affairs and Housing, Government of Ontario, *The Greenbelt Plan (2005)*, last modified January 26, 2016, http://www.mah.gov.on.ca/Page189.aspx. See also David Pond, "Ontario's Greenbelt: Growth Management, Farmland Protection, and Regime Change in Southern Ontario," *Canadian Public Policy* 35, no. 4 (December 2009): 413–432.
8. Andrea Olive, "Urban Awareness and Attitudes toward Conservation: A First Look at Canada's Cities," *Applied Geography* 54 (October 2014): 160–168.
9. "*Sanfilippodytes bertae*," *Wikipedia*, last modified March 12, 2017, https://en.wikipedia.org/wiki/Sanfilippodytes_bertae.
10. It is my hope that students who have taken my course will go on to work in this field. I know of at least eight students, among the twenty-five who completed the course, who are presently enrolled in master's programs (in environment or sustainability

management) or law programs (in environmental law).

11. Dale D. Goble, J. Michael Scott, and Frank W. Davis, "Conserving Biodiversity in Human Dominated Landscapes," in *The Endangered Species Act at Thirty*, vol. 2, *Conserving Biodiversity in Human-Dominated Landscapes*, ed. J. Michael Scott, Dale D. Goble, and Frank W. Davis (Washington, DC: Island Press, 2006): 288–290; Richard Leaky and Roger Lewin, *The Sixth Extinction: Patterns of Life and the Future of Humankind* (Toronto: Random House, 1995).

BIBLIOGRAPHY

DeStefano, Stephen, and Richard M. Degraaf. "Exploring the Ecology of Suburban Wildlife." *Frontiers in Ecology and the Environment* 1, no. 2 (March 2003): 95–101.

Goble, Dale D., J. Michael Scott, and Frank W. Davis. "Conserving Biodiversity in Human Dominated Landscapes." In *The Endangered Species Act at Thirty*, vol. 2, *Conserving Biodiversity in Human-Dominated Landscapes*, edited by J. Michael Scott, Dale D. Goble, and Frank W. Davis, 288–290. Washington, DC: Island Press, 2006.

Government of Canada. Environment Canada. Species at Risk Public Registry Office. "Recovery Strategies." Http://www.registrelep-sararegistry.gc.ca/sar/recovery/recovery_e.cfm.

Government of Canada. Statistics Canada. "Population of Census Metropolitan Areas." Last modified February 10, 2016. Http://www.statcan.gc.ca.

Government of Ontario. Ministry of Municipal Affairs and Housing. *The Greenbelt Plan* (*2005*). Last modified January 26, 2016. Http://www.mah.gov.on.ca/Page189.aspx.

Leaky, Richard, and Roger Lewin. *The Sixth Extinction: Patterns of Life and the Future of Humankind*. Toronto: Random House, 1995.

Marzluff, John M. "Fringe Conservation: A Call to Action." *Conservation Biology* 16, no. 5 (October 2002): 1175–1176.

Office of the Auditor General of Canada. *Report of the Commissioner of the Environment and Sustainable Development*, chapter 1, "Backgrounder on Biological Diversity." Ottawa: Office of the Auditor General of Canada, 2013.

Olive, Andrea. *Land, Stewardship, and Legitimacy: Endangered Species Policy in Canada and the United States*. Toronto: University of Toronto Press, 2014.

———. "Urban Awareness and Attitudes toward Conservation: A First Look at Canada's Cities." *Applied Geography* 54 (October 2014): 160–168.

Pond, David. "Ontario's Greenbelt: Growth Management, Farmland Protection, and Regime Change in Southern Ontario." *Canadian Public Policy* 35, no. 4 (December 2009): 413–432.

"*Sanfilippodytes bertae.*" *Wikipedia*. Last modified March 12, 2017. Https://en.wikipedia.org/wiki/Sanfilippodytes_bertae.

University of Toronto. "Quick Facts." Http://www.utoronto.ca/about-uoft/quickfacts.

University of Toronto. Facilities and Services. Sustainability Office. *2015–2016 Sustainability Yearbook: Highlights from St. George Campus*. Http://www.fs.utoronto.ca/sustainability-office/sustainability-yearbook-2015-16/.

Teaching Critical Food Studies in Rural North Carolina

Keely Byars-Nichols

> We need to bring sustainability into our courses and disciplines as much as we need to bring the tools of our disciplines to solving the challenges of sustainability.
>
> —Dan Phillipon, "Sustainability and the Humanities: An Extensive Pleasure"

In his 2012 essay titled "Sustainability and the Humanities: An Extensive Pleasure," Daniel Phillipon asks, "How might the humanities make the case for sustainability in the United States, how might sustainability supersede other concerns as a presiding paradigm for environmental studies, and how might those of us who do something broadly defined as 'literary and cultural studies' contribute to the creation of a more sustainable world?"[1] These challenging questions guided me as I created the first critical food studies class offered at the University of Mount Olive, the small liberal arts college where I teach in eastern, rural North Carolina. Many of my students are from the surrounding rural counties in eastern North Carolina, but we also serve many international and first-generation students and second-language learners. Regardless of their specific personal contexts, they are all aware, just by looking at the fields of shifting tobacco, cotton, and soy outside our classroom windows, that we are all in a place rooted in deep histories of agrarianism and with notable food cultures.

Further, my school has a growing and dynamic agriculture program, but we currently do not have a sustainability studies program. Thus, faculty who are not in the agriculture systems, agribusiness, and agriculture education programs currently have no structured way to engage in issues of sustainability and food studies—issues that are crucial and central to the region the school serves. However, conversations to start some sort of sustainability studies program, degree, or initiative have begun, especially as we've seen interest grow in our agriculture majors. So in spring 2015, I proposed to teach a course titled Critical Food Studies, and administration and students both in and outside the agriculture department met the idea with enthusiasm. As a humanist, I realized that my approach would be different than what would be taken in the physical and social sciences; however, I believed then and believe now that the humanities perspective is unique and valuable in fields not typically thought of as humanities-based.[2] What follows is an explanation of my theoretical framework, an analysis of my students' experiences, and a reflection on how other scholar-teachers can move forward in this field.

Eastern North Carolina, its landscape consisting of large industrial fields of monoculture, is not typically thought of as a place rooted in sustainability. During my commutes headed east from my home in Raleigh to Mount Olive, I pass hundreds of acres of conventionally grown soybeans, tobacco, and cotton. This landscape of monoculture is dotted with Monsanto trademark signs, noting the hybrid breed number of the plant being grown, and large pieces of farm equipment that spray water and chemicals onto the plants. Unsustainable eating and lifestyle realities also mark this region: eastern North Carolina, where the University of Mount Olive is located, has the highest rates of obesity, cardiovascular disease, and diabetes in the state. I notice on my commute as well that while I pass several McDonald's, Hardee's, and gas stations, I pass only one grocery store and no local farmers' markets. While I'm sure the healthy options are somewhere available in this region, they are surely not as easily or as broadly accessible as the less healthy options. This type of environment struggles to sustain a healthy, vibrant population.[3]

How could I approach teaching about issues of sustainability in such an unsustainable environment? As I began thinking through my intended course outcomes, my answer to this question was twofold: (1) by addressing, through selected readings about issues germane to the region (including hog farm pollution and genetic engineering by Monsanto), the irony of talking about sustainability in a region that struggles with sustaining the health of its people and ecology, and

(2) by focusing on each student's personal stories and relationships with food, in the hope that such attention would be an empowering solution to the problems of human and ecological health. In taking this approach, I had to acknowledge the fact that I'm an outsider to this region. While I identify as a North Carolinian and a Southerner, I am not from eastern North Carolina, and I've spent my life living in suburban and urban places, not rural ones. In fact, as one student joked, I am the stereotype of the do-gooder, Prius-driving, Whole Foods–shopping white woman who might in fact seem *most* out of place in a rural setting. I mitigated my outsider status about some of these issues by drawing attention to it (through humor and levity) and by helping students engage in questions of authenticity, region, and academic authority (through scholarship). By the end of the semester, I believe that students became more literate and in some cases fluent in the language of personal and environmental sustainability. Just as important, I also became more aware of issues that insiders of this region deal with.

Going into planning the class, I knew that I was interested in the idea of insider/outsider politics as it relates to Southern, rural identity. In his essay "Slow Food for Thought" (2008), Eric Schlosser notes, "At the moment, the majority of Americans—ordinary working people, the poor, people of color—do not have a seat at this table [of the food reform movement]. The movement for sustainable agriculture has to reckon with the simple fact that it will never be sustainable without these people." Schlosser's observation, along with my preconceptions of writers like Wendell Berry as farming "insiders" and writers like Michael Pollan as "outsiders," led me to hypothesize that farming and agriculture insiders were more effective writers to use in the classroom because, to borrow Berry's phrasing from the essay "Agricultural Solutions to Agricultural Problems" (1981), they offer rural solutions to rural problems.[4] However, what my students taught me over the duration of the class was so much more complex than my initial feeling that agricultural insiders were "good" and agricultural outsiders were "bad." They taught me that when offered the opportunity to thoughtfully and critically analyze their own experiences—rural, urban, Catholic, Jewish, black, white, mixed, etc.—they became the insider experts, while the positionalities of the writers we read became somewhat irrelevant. Thus, my chapter's argument will extend Schlosser's argument by showing how food reform and more sustainable practices and approaches can't succeed without *all* voices at the table, including the students' voices. In my analysis of my students' development of critical responses to sustainability issues related to gender, race, and region, I find that the work of teaching students productively about sustainability in this region should be transformative, reflective and analytical, and activist.

Method of Inquiry: The Course Outline

Because part of what I want to achieve in writing this chapter is to help others interested in developing a food studies class at their institutions, I will describe briefly how I organized the class. My first priority was to make central the act of writing personal narrative as a means of critical analysis. With personal narrative, students become both subject and principal investigator and thus can take more ownership over their analysis.[5] My second priority was to make the readings and perspectives interdisciplinary and both academically rigorous and accessible.

In my course introduction, I posed the question "What is 'food studies'?" The first class introduced students to the debates inherent in the field by our reading and discussion of Mark Bittman, Michael Pollan, Ricardo Salvador, and Olivier De Schutter's "How a National Food Policy Could Save Millions of Lives" (2014) and Berry's "The Pleasures of Eating" (1990). Accessible and thoughtful, these writers helped us flesh out why writers, thinkers, and activists are concerned with issues like sustainability, food culture, and identity. They also showed us how authors' points of view affect their perspectives as well as their reception by diverse readers.

In the first unit, Food and Identity, I excerpted several chapters from a wonderful collection of essays, *Food and Culture: A Reader* (3rd ed.), edited by Carol Counihan and Penny Van Esterik (2013). From a mostly social science/cultural studies perspective, these readings were very challenging for my students, but they launched discussions and provided strong models for how I wanted them to structure their first major written project: the interview analysis. The readings' topics ranged from food memories to identity (race, gender, and nationality), and the essays drew on the authors' analyses of interviews they had conducted. For their projects, my students were required to interview someone they knew and write an analysis of the interview in a format like the texts they had read, using a loose version of IMRAD format (introduction, methods, results, and discussion) and analytical concepts gleaned from the readings.

In the second unit, Food and Region, students watched four episodes of the PBS series *A Chef's Life* and listened to an interview with Vivian Howard, owner of Chef

& the Farmer, the restaurant in Kinston, North Carolina, that the show follows. To diversify further their perspectives of eastern North Carolina food culture, students also read short news articles about local brewers and tobacco farmers who are part of crop transition programs.[6] Finally, I had them read about and watch interviews with Joel Salatin of Polyface Farm in southern Virginia, as well as read Barbara Kingsolver's introduction to *Animal, Vegetable, Miracle: A Year of Food Life* (2007). These two figures are not regional to eastern North Carolina, but my primary reason for these offerings was to encourage students to think about how "fringe" or "outsider" activists can inform food movements, as both Salatin and Kingsolver take relatively extreme measures in their approaches to living sustainably. At the end of this second unit, students referred to class discussion notes, informal writing assignments, and readings to craft their second major writing project: the auto-ethnography. This assignment required students to follow specific guidelines similar to those of the first written project, including writing an introduction, discussing the theoretical framework used, and analyzing specific personal examples and experience. This model proved very effective in compelling students to do the required contextual and theoretical work they needed to do in order to analyze themselves and their own experiences critically.

In the third unit, Food and Ethics, students watched several brief videos of Temple Grandin discussing her work and the ethics of animal welfare, and watched an interview with Jonathan Safran Foer about his book *Eating Animals* (2009). They also read about local farmworkers' rights issues (including pay, medical care, and exposures to pesticides), the ethics of advertising baby formula in developing countries, and environmental justice issues like pollution created by hog farms. As they read, wrote, and talked about these texts, they explored and thought about potential community partners whom they might like to contact for their third and final major project: the community partner proposal, in which they would meet and interview representatives of an eastern North Carolina community organization, identify an organizational need, and propose a solution for the agency.

In the fourth and final unit of the semester, Food and Science, we explored the science of climate change relative to farmers and attentive to the politics surrounding this scientific issue. My students were particularly interested in the irony of the politics: that most farmers are politically conservative, even as they are being affected very concretely by an ostensibly "liberal" political issue. This irony made for lively discussions about politics, stereotypes, and unsettling expectations. We also dove into debates on genetic modification, reading about the production of papaya in Hawaii, for example.[7] Even though we addressed genetically modified organisms as a scientific issue, I, as a humanist, addressed science as a cultural production that, in this example, also involves issues of food security and cultural, national, and Indigenous sovereignty.

During Unit 4, students finalized their community partner and group configurations for Project 3, the community partner proposal. Students were grouped based on their requests and interests with a variety of nearby community partners: Student Action for Farmworkers, Wayne County Public Schools Child Nutrition Services, Carolina Farm Stewardship Association (CFSA), Common Ground of Eastern North Carolina, Hinnant Family Vineyards, Mother Earth Brewing, University of Mount Olive Dining Services, and the Feast Down East.[8] Some of the organizations I suggested, but some students proposed based on their knowledge of their communities and their interests. After being assigned to groups of three and paired with a community partner, each group was responsible for interviewing someone from the organization and identifying an organizational need. While some groups were able to conduct interviews over the phone, they also were required to make a site visit. For the group assigned to Feast Down East, it meant a visit to the organization's local flower sale; for the group partnered with Common Ground, it was a visit to the local office's urban garden; for those paired with Hinnant Family Vineyard and Mother Earth Brewery, it meant a tour of a vineyard or brewery. Often students were able to identify their organization's need as a result of their site visits, but other times they had to do some creative thinking and additional research.[9] After identifying a potential need, students wrote a proposal for a solution to the problem, grounded it in some sort of outside research, and presented the proposal to the class. Typically, the "problems" and "needs" centered on lack of publicity, public relations, funding, or other resources. For example, the group partnered with CFSA identified the problem that not enough collegiate agriculture programs were aware of the work that CFSA was doing: CFSA, they argued, had a problem with messaging more clearly to a more diverse group of potential partners. So they proposed a partnership with University of Mount Olive's Agriculture Department, wherein our university farm over the course of several years would develop some dedicated organic farmland with the help of CFSA.

The students' final exam functioned like a major project in that it asked them to reflect on key themes we'd explored throughout the semester by answering the following questions, using direct references to the texts we had read:

1. We've read about several food insiders as well as discussed the idea of food outsiders this semester. Choose one food insider and one outsider we've read or read about this semester, and compare and contrast them. Does one seem more or less reliable, believable, or effective? Why?
2. What is the most pressing scientific *or* ethical issue related to food that we've read about this semester? Why is this one most important? What should be done about it and why?
3. We've read about several food activists and writers this semester who would be considered on the fringes or margins of society. Pick one or two of these characters and explain whether you think their beliefs and practices are useful to society. Why?

This approach and course structure, in general, felt like a sound one, as I felt that it achieved my main goals to encourage critical thinking and self-reflection and engage students in their communities. Even though many of my students (in this class, nearly half) were from communities outside eastern North Carolina, I wanted to ground the meaningfulness of this work in the community where they reside for most of the year: Mount Olive, North Carolina. Even if the students didn't call this region home, I think that the approach still resonated with them because it made them more aware of, and more connected to, the culture of the region.

Results and Analyses

As I stated in my introduction, my guiding tenets in teaching about sustainability are that the work should be transformative, reflective and analytical, and activist. *Transformative* work means that the student's perspectives and values get refined, though not necessarily changed. Transformation, which is related to deeper understanding, is central to the work I do—specifically in the writing and literature classroom, and generally as a humanist teacher-scholar. In this work, I am concerned with interpreting and making meaning out of texts, experiences, and acts, and in this class, too, I crafted assignments and projects to push students to some sort of transformation of self, including deeper understanding of self, others, and reality. I saw the most transformation in students' thinking in their interview projects (Project 1, described above).[10] In the two student samples I offer below, students interviewed parents and arrived at a deeper understanding of their own identities. Of his identity, John writes:

> My mom has always had a special way of preparing food ever since I was a kid; being raised in a Jewish household isn't like any other culture. I have always been curious about food preparation in the Jewish community but have never taken the time to talk to my Mom about it or actually learn. . . . I learned a lot from the interview and have a totally different perspective towards the way we eat. . . . My mom explained to me the ways of Judaism and how she would follow the way her grandfather ate by not eating dairy with meat, and how certain parts of animals are not to be eaten. . . . I learned way more than I expected when going into this essay. . . . My food beliefs have changed.

In this excerpt, John describes with thoughtful detail the history behind practices that he had always observed, but had not fully understood. He bears witness to an internal change in his thinking brought about by conducting the interview when he concludes, "I have a whole new perspective on the way my Mom raised me and I would like to take on the Jewish dietary laws." My goal of pushing students to experience a transformation either through change or refinement of beliefs is exemplified here.

Another student experiences a similar shift when she completes an interview with her father. Generally, her shift is around the way she thinks about her father as sole provider and preparer of all her family's meals. Mary writes:

> Having grown up in rough conditions, I quickly became familiar with the true value of a dollar. My two brothers, mother, father, and I lived off of paycheck to paycheck[;] therefore receiving one of the name brand boxed cereals rather than the store brand kind was the best day of the month. [Then] my father [became] the sole provider after my mother passed away. . . . [During this interview] I was genuinely breathless with some of my father's answers because I expected the exact opposite from him. You never know what a person is capable of. . . . You can get to know a person through the way they cook and what they cook. After listening to my father talk about his childhood and the amount of effort he puts into feeding my brothers and I with the best food out there, I realized that he does so much more for us than I tend to notice. . . . I always use[d] to think that he just put on a fake smile to hide the real annoyance he had by being the sole chef of the house.

> After conducting this interview I can definitely say that I value food a lot more now. . . . Most importantly, make sure to thank those who have passed their traditions down to you and continue to pass them on to fellow generations.

The evident tenderness with which this student reflects upon her family history emphasizes the shift that she describes in her thinking about her family's food practices. When she realizes her father's hard work and joy related to cooking, I see a generational shift. This student's thinking about her father has changed: she now sees the intentional decisions he made in raising and feeding his children, and she seems to have internalized her understandings in a way that she will likely pass on to "fellow generations." In her father, she sees that hard work can also be something that is deeply fulfilling and satisfying. Mary's thinking is transformed as she sees the past more clearly by having this conversation with her father, and she understands more deeply the effect that her family's food culture has on her now and in the future.

Another assignment that showcased students' transformation was their reflection on the community partner proposal, a group project. The two types of transformative thinking I observed were social and professional. As often happens, students who never had conducted meaningful group work before seemed amazed that it could be useful. In particular, the group who visited a local vineyard had an especially meaningful experience working together. One student, Jane, in this group writes:

> The best part of this project was that I was working with new people and sometimes it is hard to talk to new people, and [get] them involved. . . . [But we] literally sat outside on the steps of the library on a Friday night giggling away to ourselves, trying to finish our presentation. . . . A whole new group of people that I had never talked to before I came into this class and now they're people I will talk to around campus. They are easy to work with[;] we all had opinions to put forward, and our ideas were put together to create what we had[;] we all sat together, and it was actually fun.
>
> Our biggest success was working together as a team. . . . We found ourselves more engaged in the tour and discussions following as opposed to other projects we have little interest in.

Although group work is always challenging to design and implement, it clearly worked very well on an interpersonal level in this group. This group's responses reflect growth and deeper understanding of the interpersonal skills required and the social gains that can be earned through productive work with others in one's community.

Similarly, I saw students come to deeper, transformative understandings of their professional goals in these group projects. One student came closer to naming and identifying a career goal that he never knew had a formal educational path: beer brewing. "The whole way through my biggest strength with this project was giving background on brewing beer and the process of a working sustainable brewery," he writes. "This project has played a large role in helping me make some important decisions for my future. I am considering moving into the microbrewing field." This sort of realization, where personal interests are transformed into actionable professional plans and discourses, are perhaps the rewarding teaching moments that most powerfully motivate what we do in our classrooms. In order to make education more sustaining for our students, especially for students who are entering the work force in increasingly unsustainable environments, these sorts of transformational educational experiences are crucial.

As I discussed earlier, my focus on personal narrative was central to the work I wanted to do in this class. My rationale for this approach is based on what my students teach me in the composition classes I instruct: I find that when they speak, research, and write about topics in which they are "experts," they develop a more authoritative voice and arrive at more thoughtful analyses. When I ask students to write about themselves, I often use the word "reflect" and ask questions like "What do you think that event meant to you then and now?" The word "reflect" seems to make more sense to them than the word "analyze," even though the latter is what they're doing when they answer my reflective questions. They are making meaning out of an event that happened to them; they are analyzing and interpreting in their own voice. As Jane Danielewicz writes in "Personal Genres, Public Voices" (2008), "Text with voice exhibits authority, the weight or totality of a person's presence of a believable constructed persona behind the words. This element of agency—that writing is action and that voice increases its power—is what makes voice such a crucial quality." She asserts that finding this personal voice and agency allows students to move into public debates with a "public voice," "one that enters the ongoing conversation to change, amend, intervene, extend, disrupt, or influence it."[11] Thus, a public voice can create meaningful analysis and critical thought.

I heard my students honing their public voices most effectively in Project 2, the auto-ethnography. In the first example, I witnessed a student, Henry, reflecting on his personal connection to eastern North Carolina barbecue traditions, as well as contextualizing and analyzing that experience's larger meaning:

> Many times growing up and even to this day I will volunteer to help my grandpapa cook the pig whatever the occasion.... The time taken to cook is not only important to cooking BBQ but the possible application of virtues, such as patience is an essential ingredient to preparing many culinary dishes across many ethnic groups. Time must be taken both for safety and for flavor and in the case of BBQ in Eastern North Carolina.... Food itself is not the topic here but rather the atmosphere in which one's life is set in. It is a bond that I'm sure affected more than my food identity in a positive way. The time that is spent joking and telling about each other's day may not seem like much at first, but communication is the key to forming family relationships. The fact that it is a family dinner also has an effect on the meals that are prepared and the quantity prepared.... It has always been a part of my culinary atmosphere that it will revolve around family and sharing the meal with them.

This student also provided secondary research on the history of eastern North Carolina barbecue traditions to supplement his personal narrative. He carefully parallels his own experiences to academic research to show that his experiences align with a larger, older, regional food tradition. By examining his own personal experience, the student arrives at a thoughtful and sophisticated argument in line with what we see when writers like Berry and Kingsolver explore the pleasure of eating locally. Despite the fact that he does not acknowledge that the pork he grew up enjoying and deciding to write about results from unsustainable production practices, this student's response zeroes in on a problem Kingsolver observes.[12] Eating locally often means not eating organically or sustainably, perhaps especially in eastern North Carolina, where sustainable food practices currently face too many formidable economic, educational, and political factors to flourish.

The use of the public voice to analyze personal narratives leads me to my next point: activism. Despite the opinions of critics of any higher education project that engages students in political issues, I find such action a primary responsibility of higher education. Indeed, many more scholars have shown us how introducing students to the politics of power and social justice can work toward clearer critical thinking and writing and more productive working conditions. In "Personal Experience Narrative and Public Debate: Writing the Wrongs of Welfare" (2008), Lorrain D. Higgins and Lisa D. Brush explain, "Activist rhetoric is an expanding area of scholarship and teaching that includes service learning, community-based writing projects, and the publication of action research in which researchers not only observe but intervene in participants' literate practices to promote social justice." This type of rhetoric is precisely what my assignments urged students to approach, specifically as they engaged with community partners. Indeed, the field of sustainability studies is uniquely positioned to encourage such work. Janet Fiskio writes, "From its inception, the field of American ecocriticism"—an iteration of sustainability studies, I contend—"has been fixated on the reinhabitation of the local community as a means of simultaneously regenerating ecological sensibility and participatory democracy."[13] During the Critical Food Studies course, students engaged in many inherently political issues: farmworkers' rights, organic farming, food sovereignty, environmental justice, animal welfare, food security/scarcity in rural America, and proper nutrition in public schools. I was not, however, hoping to indoctrinate students into any progressive or conservative political agenda that might be associated with activism surrounding these issues. Rather, I wanted them to learn about the multiple perspectives relating to these issues and figure out how such issues might affect their lives and communities.

In the community partner proposals, for example, students were asked to imagine what they could do to solve an organization's concern or need. One group, proposing a partnership between our university and CFSA, explained their plan in this way:

> Our proposal is to create a partnership between the University of Mount Olive and Carolina Farm Stewardship Association (CFSA). CFSA is an organization that advocates for fair farming and food policies while building systems and educating communities on local, organic farming. With this mutual agreement, our goal is to develop a self-sustainable organic farm that provides various types of produce for those affiliated with the University of Mount Olive and the surrounding community. After establishing the Campus Farmer's Market, we would like to expand the quantity of our production to supply the dining services on campus. . . . The transition from traditional methods to organic methods takes about three to five years. . . . In developing this partnership with CFSA we will be providing a service for the University of Mount Olive and the local community.

Here, the students are not simply endorsing the value of organic farming and food system reform or eschewing the evils of conventional farming; they are imagining how this organization could make their community more productive and sustainable. Similarly, another group proposed a partnership with Student Action for Farmworkers, a national organization with many local chapters that works to engage student groups in farmworker advocacy:

> [The Student Action with Farmworkers] organization tries to reach out to all students around North Carolina and have them volunteer and help the farmer.... They believe that raising awareness will help stop the horrible things being done. The main issue is not enough people know what is going on at farms because not many people volunteer or think about where [their] food is even coming from. A solution to this is to bring more awareness around.... It would be great to have a fundraiser at UMO in order to show all the people here that farmers aren't treated right. Our students are the new future and if we care more about farmers' rights so will all of society.

In both of these groups, the students' personal voices become part of public, political discourse, as they imagine what their theoretical proposals could do if enacted.[14]

In the students' final exam responses, I continued to see the emergence of their activist voices. Regarding the issue of vegetarianism, most students provided thoughtful and complicated responses to the final exam question, which asked them to write about one or more of the food activists and writers we had studied during the semester who might be considered on the fringes or margins of society. Students were to explain clearly whether they thought the beliefs and practices of the activists or writers were useful to society, and to explain their opinions. Instead of declaring a false and perhaps extreme promise to become a vegetarian, students considered their own beliefs seriously, often suggesting increased government regulations that would improve the lives of animals we eat. For example, Caroline writes, "I consume meat so I cannot say that we should stop slaughtering animals and all become vegetarians." Yet she believes "that farm animals [should be] treated with the respect they deserve" and that "without activists like Foer, people would be clueless about the multiple issues that farming has on the environment, animal rights, and global warming." Indeed, very few students expressed the intent to make different choices by, for example, deciding to eat less meat, boycott certain companies, shop and consume differently, or join political action groups. However, their shift in thinking about who is responsible for unsustainable and toxic environmental problems (in the above example, Caroline points to a lack of effective government regulations) is to my mind very significant because it shows that they have begun to imagine that their choices *do* and *can* make a difference.

To this same exam question, another student, Chelsey, used the example of pollution caused by hog farms to express her new political views about environmental justice. She writes:

> I believe the most ethical issue related to food that we have discussed this semester was the pollution of hog farms and how it was affecting the environment. In North Carolina, residents everywhere are affected from the pollution of pig farmers because they are not disposing of pig waste/feces properly. It is polluting water and killing fish in the Neuse River Basin. . . . This ethical issue is extremely important because it affects the environment and the people around the area of where it is occurring; and something can be done about it, yet it is not happening. The state needs to come to an agreement with farmers and their hog farms on how to properly dispose of the hog waste. No amount of money for the technology should matter because it is costing lives, so this should not even be a problem at all.

Again, I hear the student suggesting that this is both an ethical and a political issue, not just a business issue. As Chelsey reflects on the extent and severity of the problem, as explored in our class readings and discussions, it seems obvious to the student that political action must be taken. While her conclusion that "this should not even be a problem at all" might seem naive, her statement acknowledges that the hog farming industry has the technology and profit margin to make changes and that the legislature has the ability to increase regulation, but that both parties are choosing otherwise. In UMO's five-county service region, for example, concentrated animal feeding operations (CAFOs) produce 15.5 million tons of hog manure a year.[15] This extreme volume, combined with a lack of effective regulations, has resulted in unhealthy conditions in many of the communities where these CAFOs are located.

In these and other examples, students reflect a nascent awareness that "democracy is built on the ideal of widespread participation, on the principle that a broad spectrum of citizens can deliberate about shared problems and possibilities." This consciousness is perhaps especially important because many students view themselves as subordinate subjects, either because they feel marginalized due to race, gender, or class, or because they don't feel part of the system of higher education because of their first-generation status and/or age. As Higgins and Brush further remind us, subordinated groups include individuals to whom the democratic process is most crucially important, and "to assume that subordinated groups should be shielded from engaging difficult questions and assumptions or that they must be protected from [them] is patronizing."[16]

Reflection

In reflecting on this teaching experience, I would change very few things. While the group projects were meaningful, these sorts of projects are always very complex. To enhance group work with community partners, I would set up relationships beforehand and then make individual group members more visibly accountable by, for example, having students copy me in every correspondence with one another and their community partner. What's more, I would team-teach with a faculty member from the agriculture sciences. Planning-time constraints prevented me from doing so the first time, but I can see many benefits from setting up this course as an interdisciplinary one with an expert from the humanities and an expert from the physical sciences. Likewise, Phillipon underscores the need for more interdisciplinary conversations in sustainability studies:

> While we in the humanities certainly need to make a place for ourselves at the sustainability table, we also need to collaborate more with our colleagues in the natural and social sciences, recognizing that interdisciplinarity goes both ways. We need to be realistic about the differences between disciplinary cultures, of course, including differences within humanities themselves, but if any of us see ourselves first as disciplinary representatives, we haven't gotten the message.[17]

Here, Phillipon advocates an interdisciplinary approach that goes beyond the "marching through the disciplines" that often accompanies interdisciplinary pedagogical approaches, and he urges scholar-teachers to engrain interdisciplinary approaches into our own practices. The field of sustainability studies encourages and often requires this sort of unsiloed approach. Finally, I would make a class like this part of a two-semester series that involves actually implementing the ideas the students proposed in their group proposal projects. A service-learning project like collaborating with a community partner is perhaps one of the best ways to build sustainability into programs by making the students work to support the campus and the larger community that sustains the university. Indeed, if our fields and the project of the university are to remain sustainable, new approaches and methodologies are the key to our survival.

My personal journey in teaching this course has also led me to a greater understanding of the complicated concerns of the region. For example, my personal hard line against Monsanto—a company that my colleagues in agricultural science endorse and on which my students who grew up in farming families rely for their livelihoods—will shut down dialogue. I learned instead to keep my awareness of my outsider status at the forefront in my conversations about such controversial topics and to listen more closely before relying on any automatic or partisan-seeming response. Further, I have learned to think differently about what a campus like the University of Mount Olive can do to participate in educational, personal, economic, and environmental transformation. The type of student-activist thinking that I bore witness to during this semester filled me with ideas for real action that students could engage in if the course were taught as a two-semester project. Over time, this sort of community engagement could do so much to affect the health and environment of the entire region. Instead of regarding the institution as a stronghold of conservatism that is reluctant to change, especially when that change may align with more liberal political positions (organic farming and increased government regulations, for example), I can now envision many ways that the university can do more to sustain this region and its people.

NOTES

1. Daniel J. Phillipon, "Sustainability and the Humanities: An Extensive Pleasure," in "Sustainability in America," special issue, *American Literary History* 24, no. 1 (Spring 2012): 163. In this chapter, I sometimes will use the terms "critical food studies" and "sustainability studies" interchangeably because I see the former as a subset of the latter.
2. Indeed, current research in this area urges scholars to cross disciplinary lines to engage in more productive and meaningful thought and action surrounding issues of sustainability in higher education. Phillipon touches on this when he suggests that humanists "need to collaborate more with our colleagues in the natural and social sciences, recognizing that interdisciplinarity goes both ways" (Phillipon, "Sustainability and the Humanities," 169). Gillen D'Arcy Wood adds to this sentiment when he states that the humanist perspective is crucial because the "quantitative approach to sustainability only goes so far." However, he continues, "Sustainability studies, for humanists, in turn mandates scientific literacy, a comfort with quantitative methods, and active engagement and collaboration with

scientists, social scientists, and policymakers across the sustainability arena" ("What Is Sustainability Studies?" in "Sustainability in America," special issue, *American Literary History* 24, no. 1 [Spring 2012]: 4, 9).

3. "About the Center for Health Disparities," East Carolina University, Brody School of Medicine, Center for Health Disparities, last modified January 27, 2016, http://www.ecu.edu/cs-dhs/healthdisparities.
4. Eric Schlosser, "Slow Food for Thought," *The Nation*, September 22, 2008, 5; Wendell Berry, "Agricultural Solutions to Agricultural Problems," in *The Gift of Good Land: Further Essays Cultural and Agricultural* (1981; Berkeley, CA: Counterpoint, 2009), 113–124.
5. See Peter Elbow, "Can Personal Expressive Writing Do the Work of Academic Writing?" in *Everyone Can Write: Essays toward a Hopeful Theory of Writing and Teaching Writing*, ed. Peter Elbow (New York: Oxford University Press, 2000), 315–318.
6. The Tobacco Transition Payment Program (TTPP) began in 2004 with the end of all regulation of U.S.-grown tobacco, a shift that effectively propped up the waning industry. With the end of these quotas, the U.S. Department of Agriculture began buying out tobacco farms. Most tobacco farmers used the lump sums to transition to other crops or retire, as demand continued to shrink. See Blake Brown, "The End of the Tobacco Transition Payment Program" (Raleigh: North Carolina State University, November 14, 2014), https://tobacco.ces.ncsu.edu/wp-content/uploads/2013/11/The-End-of-the-Tobacco-Transition-Payment-Program.pdf?fwd=no.
7. Amy Harmon, "A Lonely Quest for Facts on Genetically Modified Crops," *New York Times*, January 14, 2014.
8. Student Action for Farmworkers is an organization that works with farmworkers, students, and other advocates primarily in the southern United States to make farmworkers' working conditions safer, more equitable, and more just. The organization mentors, educates, and mobilizes all levels of advocates of this issue. For more information, visit https://www.saf-unite.org.

   The Carolina Farm Stewardship Association (CFSA) is a nonprofit organization that helps people in North and South Carolina "grow and eat local, organic foods by advocating for fair farm and food policies, building the systems family farms need to thrive, and educating communities about local, organic agriculture." The association does so by creating partnerships between farmers and businesses and by helping farmers transition to organic practices. For more information visit http://www.carolinafarmstewards.org.

   Common Ground Eastern North Carolina is a small grassroots group in Kinston, North Carolina, that operates a small city farm for the purposes of increasing healthy food supplies, educating, and advocating for the community. For more information, visit https://www.commongroundenc.com.

Wilmington, North Carolina–based Feast Down East helps smaller farmers and growers find and maintain markets. For more information, visit www.feastdowneast.org.

9. For example, the group who visited Hinnant Family Vineyards realized on their site visit that the family-owned business might not be maximizing profits by fully using all production capacity and retail space available to them. Several business majors in the group were able to guide the group in drafting a business proposal for expansion that required relatively few costs and potentially significant increased revenues. Alternatively, the group who visited Mother Earth Brewing proposed a partnership between UMO and the brewery based on their findings of the growing popularity of fermentation studies as a major at several North Carolina universities.
10. For the purposes of this chapter, I use the word *transformative* in a way similar to the way Allison Cook-Sather uses the word *translation* in that she explains that "translation" (of thought into written word, memory into analytical narrative) is an act that makes a text "more vital, richer, more resonant, and more open to expression and to interpretation" ("Education as Translation: Students Transforming Notions of Narrative and Self," *College Composition and Communication* 55, no. 1 [September 2003]: 95).
11. Jane Danielewicz, "Personal Genres, Public Voices," *College Composition and Communication* 59, no. 3 (February 2008): 425.
12. See Wendell Berry, "The Pleasures of Eating," in *What Are People For?* (New York: North Point Press, 1990); Barbara Kingsolver, *Animal, Vegetable, Miracle: A Year of Food Life* (New York: Harper Collins, 2007). Later in the semester, we addressed the issue of industrial pork production, especially as it relates to eastern North Carolina politics. We read and discussed Dana Fine Maron, "Defecation Nation: Pig Wastes Likely to Rise in U.S. Business Deal" (*Scientific American*, July 12, 2013, http://www.scientificamerican.com/article/smithfield-pig-waste/), detailing the practices of Smithfield Foods Inc. (a company based in eastern North Carolina) that contribute to serious health and environmental risks for communities living near hog farms and pork production facilities.
13. Lorraine D. Higgins and Lisa D. Brush, "Personal Experience Narrative and Public Debate: Writing the Wrongs of Welfare," *College Composition and Communication* 57, no. 4 (June 2006): 724; Janet Fiskio, "Unsettling Ecocriticism: Rethinking Agrarianism, Place, and Citizenship," *American Literature* 84, no. 2 (June 2012): 301; see also Stanley Fish, *Save the World on Your Own Time* (New York: Oxford University Press, 2008); Paulo Freire, *Pedagogy of the Oppressed*, trans. M. B. Ramos (New York: Continuum, 1989); Ira

Shor, *Empowering Education: Critical Teaching for Social Change* (Chicago: University of Chicago Press, 1992).

14. As exercises in creative and critical thinking, students were unable to implement these plans and proposals because of time limitations. In the future, I would like to teach the class as a two-semester series, so students could work through some of the proposals.
15. U.S. Government Accountability Office, *Concentrated Animal Feeding Operations: EPA Needs More Information and a Clearly Defined Strategy to Protect Air and Water Quality from Pollutants of Concern* (Washington, DC: U.S. Government Accountability Office, September 2008), http://www.gao.gov/new.items/d08944.pdf.
16. Higgins and Brush, "Personal Experience," 695, 716.
17. Phillipon, "Sustainability and the Humanities," 169.

All student names have been changed. I wish to thank all members of the spring 2015 offering of this course for their contributions to the thoughts developed in this essay, particularly Gina Fader, Brian Stewart, Benedetta Abbate, and Davyd Powell.

BIBLIOGRAPHY

Berry, Wendell. "Agricultural Solutions to Agricultural Problems." In *The Gift of Good Land*, edited by Wendell Berry, 113–124. Berkeley, CA: Counterpoint, 2009.

———. "The Pleasures of Eating." In *What Are People For?* edited by Wendell Berry, 145–152. New York: North Point Press, 1990.

Bittman, Mark, Michael Pollan, Ricardo Salvador, and Olivier De Schutter. "How a National Food Policy Could Save Millions of Lives." *Washington Post*, November 7, 2014.

Cook-Sather, Allison. "Education as Translation: Students Transforming Notions of Narrative and Self." *College Composition and Communication* 55, no. 1 (September 2003): 91–114.

Counihan, Carol, and Penny Van Esterik, eds. *Food and Culture: A Reader*. 3rd ed. New York: Routledge, 2013.

Danielewicz, Jane. "Personal Genres, Public Voices." *College Composition and Communication* 59, no. 3 (February 2008): 420–450.

Elbow, Peter. "Can Personal Expressive Writing Do the Work of Academic Writing?" In *Everyone Can Write: Essays toward a Hopeful Theory of Writing and Teaching Writing*, edited by Peter Elbow, 315–318. New York: Oxford University Press, 2000.

Fish, Stanley. *Save the World on Your Own Time*. New York: Oxford University Press, 2008.

Fiskio, Janet. "Unsettling Ecocriticism: Rethinking Agrarianism, Place, and Citizenship." *American Literature* 84, no. 2 (June 2012): 301–325.

Foer, Jonathan Safran. *Eating Animals*. New York: Little, Brown, 2009.

Freire, Paulo. *Pedagogy of the Oppressed*. Translated by M. B. Ramos. New York: Continuum, 1989.

Higgins, Lorrain D., and Lisa D. Brush. "Personal Experience Narrative and Public Debate: Writing the Wrongs of Welfare." *College Composition and Communication* 57, no. 4 (June 2006): 694–729.

Kingsolver, Barbara. *Animal, Vegetable, Miracle: A Year of Food Life*. New York: Harper Collins, 2007.

Phillipon, Daniel J. "Sustainability and the Humanities: An Extensive Pleasure." In "Sustainability in America." Special issue, *American Literary History* 24, no. 1 (Spring 2012): 163–179.

Schlosser, Eric. "Slow Food for Thought." *The Nation*, September 22, 2008. Https://www.thenation.com/article/slow-food-thought/.

Shor, Ira. *Empowering Education: Critical Teaching for Social Change*. Chicago: University of Chicago Press, 1992.

Wood, Gillen D'Arcy. "What Is Sustainability Studies?" In "Sustainability in America." Special issue, *American Literary History* 24, no. 1 (Spring 2012): 1–15.

PART 3.

Reinhabiting and Restoring Who and Where We Are

Mindfulness, Sustainability, and the Power of Personal Practice

Jesse Curran

> We don't have to sink into despair about global warming; we can act. If we just sign a petition and forget about it, it won't help much. Urgent action must be taken at the individual and collective levels. We all have a great desire to be able to live in peace and to have environmental sustainability. What most of us don't yet have are concrete ways of making our commitment to sustainable living a reality in our daily lives. We haven't organized ourselves. We can't simply blame our governments and corporations for the chemicals that pollute our drinking water, for the violence in our neighborhoods, for the wars that destroy so many lives. It's time for each of us to wake up and take action in our lives.
>
> —Thich Nhat Hanh, "The Bells of Mindfulness"

Adjunct professor. PhD in English, specialization in environmental poetics. Affiliated with departments or programs in English, sustainability studies, writing, honors, and first-year experience.

Yet when it comes to identifying as a practitioner of yoga, I've been told to think twice.

I am an adjunct professor at Stony Brook University, the State University of New York's flagship STEM research institution. I earned my PhD in English at Stony

Brook in 2012, exploring meditation as a mode of reading and interpreting poetry with breath, concentration, and ecological praxis. Since my funding ran out in 2010, I have been employed in a contingent capacity teaching in numerous departments and programs. And when I recently prepared my academic job application letter, a senior faculty adviser counseled me not to use the word "yoga."

The adviser's intentions were sincere enough; he thought that the word might estrange certain readers and inhibit my chances at getting interviews for ever-elusive tenure track positions. Although I declined to take it, I was deeply troubled by this advice, as it seemed to touch the heart of something direly unsustainable about academic practice. Candidates are encouraged to mask foundational components of their identities as teachers and thinkers in order to play more successfully into a market that ultimately seems to alienate them from their sources of strength. What this colleague's comment reminded me of is a tendency in academia to compartmentalize identities and to dissociate from integration. At times, this compartmentalization can be manifested in the gulfs between speech and action, between intellect and emotion, and, perhaps most critically for sustainability, between theory and practice. And although these days the word "yoga" might bring to mind a stereotype of spandex pants, tight abs, and suburban housewives, it is an ancient form of meditation and practice of mindfulness that harbors a rich intellectual tradition and offers highly practical tools for engaging with the world's problems.

In my experience, the practice of yoga and meditation has offered me, to echo Thich Nhat Hanh, more concrete ways of making a commitment to sustainable living in my daily life. In 2012, I completed a teacher training in classical yoga, and I have taught a weekly hatha yoga class at a local studio since. I also continue to study texts such as the *Bhagavad Gita* and Patañjali's *Yoga Sutras*, as well as reading the work of contemporary teachers such as B. K. S. Iyengar and Stephen Cope. I came to hatha yoga early in my PhD studies, and it has transformed my pedagogy insomuch that alongside teaching intellectual development and specific subject matter, I aspire to provide students with the tools that enable them to "wake up and take action" in their own lives. For these reasons, and as I hope this chapter explains, I proudly keep the word "yoga" in my professional materials, as I feel it is appropriate and valuable in guiding students to understand and act upon the questions put forth by contemporary sustainability studies.

At its root, sustainability suggests endurance, deriving from the Latin *sustinere*, to hold up, hold upright; furnish with means of support; bear, undergo, endure.

These roots suggest the challenge that sustainability presents: to bear and endure, one needs to be equipped with strength, commitment, and discipline. Indeed, the statistics and dialogues that surround climate change are heavy weights to bear, so much so that for many, it seems easier to look the other way and continue to maintain the status quo. Whereas higher education for many has become less personally and professionally sustainable, this era of climate change has seen "sustainability" amplified as a concept, theory, movement, buzzword, and marketing agenda. In *Sustainability Principles and Practice* (2014), Margaret Robertson defines it as a perceptive lens:

> Sustainability is about seeing and recognizing the dynamic, cyclical, and interdependent nature of all parts and pieces of life on earth, from the soil under our feet to the whole planet we call home, from the interactions of humans with their habitats and each other to the invisible chemical cycles that have been redistributing water, oxygen, carbon, and nitrogen for millions of years.

She frames sustainability through present-participle verbs and locates it as a process of "becoming educated and involved citizens of this living, changing world and determining what most needs to be done and which part each of us will take in our individual corner of the world."[1] Concerned with "becoming" and "determining," sustainability is grounded in process orientation, which facilitates more educated involvement with one's environmental system. However, the very nature of process orientation resists the product-driven outcomes that often frame mainstream educational models.

The mindfulness that is cultivated by yoga's meditative practices can help to yoke education with involvement, while also facilitating a philosophical worldview based on the dynamic, interdependent nature of life on earth; yoga's foundational impulse, put simply, is integrative. Yoga also can generate strength, discipline, and endurance in both mind and body, and it can provide individuals with practical means of bearing the weight of sustainability. The integration of such processes into the higher education classroom, while not easily quantifiable, nonetheless helps students develop and accentuate skills that can directly address core challenges of sustainability. This integration is especially important in light of the mainstream educational model that, in the words of David W. Orr, "toward the natural world . . . emphasizes theories, not values; abstraction rather than consciousness; neat answers instead of questions; and technical efficiency over conscience."[2]

Not all intellectuals have been hesitant to identify themselves with yoga. In his journal, Henry David Thoreau famously wrote:

> Depend upon it that, rude and careless as I am, I would fain practice the *yoga* faithfully. The yogi, absorbed in Contemplation, contributes in his degree to creation: he breathes a divine perfume, he hears wonderful things. Divine forms traverse him without tearing him, and, united to the nature which is proper to him, he goes, he acts as animating original matter. To some extent, and at rare intervals, even I am a yogi.

Thoreau's yoga had nothing to do with our current and popular, primarily physical version of yoga; his sense derives from Charles Wilkins's translation of the *Bhagavad Gita*, which he was reading during his time at Walden Pond. As he writes in *Walden* (1854),

> In the morning I bathe my intellect in the stupendous and cosmogonal philosophy of the Bhagvat-Geeta, since whose composition years of the gods have elapsed, and in comparison with which our modern world and its literature seem puny and trivial; and I doubt if that philosophy is not to be referred to a previous state of existence, so remote is its sublimity from our conceptions.[3]

Yogic philosophy as presented in the *Bhagavad Gita* is built upon an understanding of the interconnectedness of all existence, implying ethical precepts based on mindfulness practices, many of which are recurring themes within Thoreau's didacticism.

Thoreau has been hailed as an icon of American environmentalism, and his insistence on contemplative action has since been foundational to movements for both social and environmental justice. My admiration for Thoreau as a public intellectual, an environmentalist, and a writer runs deep; his ecological-meditative poetics were a central subject in my dissertation research. In fact, I dedicate an entire chapter to revealing how literary metaphor in *Walden* functions to bring readers through a process of mourning that results in a new state of ecological being. Thoreau's particular power lies in his rhetorical position, through which he impels readers to examine their own situations and choices. As Stanley Cavell writes of *Walden*, "It would be a fair summary of the book's motive to say that it invites us to take an interest in our own lives, and teaches us how."[4]

Also noting the ways in which *Walden* instructs self-evaluative practices, E. B. White has written, "If our colleges and universities were alert, they would present a cheap pocket edition of the book to every senior upon graduating, along with [a] sheepskin or instead of it." In light of climate change, White's commentary on *Walden* and higher education is both relevant and telling; when read carefully, students cannot help but be challenged by Thoreau's call to action, and I have found that my students consistently respond to his confrontational tone in "Economy." Thoreau's cutting examination of consumption and personal responsibility, combined with a rhetorical urgency that confronts his readers, make his eerily prophetic text just as relevant as ever. Indeed, Thoreau is his own sort of yogi, and in a way, *Walden* invites its readers to practice their own forms of meditation. Ultimately, the yogi is one who works to bring her theories into practice. Or, as Thoreau famously wrote, the philosopher is one who seeks to "solve some of the problems of life, not only theoretically, but practically."[5]

Remembering to Be Tall

Like Thoreau and despite my adviser's well-meaning advice to the contrary, I openly identify myself as a practitioner of yoga because it has helped me persevere. After the collapse of a long-term relationship, the disillusionment of the looming job market, and the necessity of relying on part-time work to finish my degree, I found myself truly living an unsustainable academic lifestyle. I was in graduate school on Long Island—a place with one of the highest costs of living in America—and received four years of funding, even though I needed seven years to complete my doctorate. As a result, like so many of my peers, I took work as an adjunct to make up the difference. My salary was $2,500 for a full semester class, in a place where the monthly rent for my one-bedroom apartment was $1,250. I found myself internalizing Stony Brook's pitfalls, particularly the way in which the corporate university system strips away individual identities, reducing human beings to numbers and symbols (usually dollar signs), in combination with my own wavering self-esteem—and a depressive anxiety began to haunt my days.

At one point, as I was contemplating the challenges ahead of me, I considered leaving my program. But my dissertation adviser offered me some personal advice. He said to me frankly, "Jesse, remember that you're tall." I took the word literally. Tall. I needed to stop hunching forward and drawing inward. I needed to dedicate time to my physical posture so I might fortify my emotional landscape. I needed to stand upright and draw my shoulder blades together. And so, I found *tadasana*, mountain pose. An essential and foundational yoga posture, *tadasana* is the literal manifestation of groundedness and personal conviction. If I was going to have the strength to see my degree and career in academia through and be a writer with a strong voice willing to take risks and shape her own sense of truth, I needed to integrate my physical and emotional strength to hold my posture in both body and mind.

I turned to the practice of hatha yoga in order to release the mental and emotional tension that had rooted its way into my body. What came with the practice was strength in body, mind, and spirit—and an awareness of the intrinsic connectedness of these facets of my identity. My path of hatha yoga also concerns care, discipline, and widened perspective. Most significantly, perhaps, it helps me to establish and articulate my ethical compass, which manifests in an ongoing attempt to internalize the *Gita*'s most penetrating and repeated lesson: action without attachment to the fruit of action. I completed my doctoral studies by understanding them to be a process-oriented pursuit of knowledge essential to my development as a citizen and community member, rather than a product-oriented action toward a diploma, honor, or job. In the morning, just hours before I defended my dissertation, I woke early to practice *tadasana*, a pose I held throughout my defense. I was able to stand up and defend my dissertation confidently because of this pose, and my research on ecological poetics was profoundly shaped in just the same way.

In the spirit of both Hanh and Thoreau, my conviction that mindfulness and meditation are major assets to current sustainability studies (and education writ large) lies at the heart of my research, my pedagogy, and my existence as a member of an ecological community. Is education merely a transference of facts (the mutability of which are staggeringly volatile), or does education imply a skill set or methodology for processing stimuli in many myriad forms? Is there an ethic at work here? Philosophers since time immemorial have purported that understanding must, as a matter of course, begin with the self, and without that foundation to build upon, all other "learning" must be seen as suspect. It is for this reason, mainly, that I do not mask my allegiance to the ecological virtue ethics as voiced by writers such as Aldo Leopold, Rachel Carson, Wendell Berry, Orr, Gary Snyder, and Terry Tempest Williams, and I have found that yogic meditation and mindfulness can become a means of bringing these ethics into fruition. As Snyder notes, "The yogin is an experimenter whose work brings forth a different sort of discourse, one of deep hearing and doing. The yogin experiments on herself. Yoga, from the root yaj (related to the English 'yoke'), means to be at work, engaged."[6] The practice of yoga allows individuals to create personal strength and perceptive clarity, both of which are necessary in order to follow the ethical precepts that the practice opens.

But what more specifically do I learn from my yoga practice, and how might those lessons translate into an ecological pedagogy? Perhaps most importantly, my practice reveals the versatility of the "beginner's mind," which is a seminal concept in both yogic and Zen practices and works with distinct grace when encouraging students to approach challenging ideas that ask them to radically examine their own behaviors. As Shunryu Suzuki explains, a beginner's mind asks individuals to become aware of preconceptions and strive toward openness: "If your mind is empty, it is always ready for anything; it is open to everything. In the beginner's mind there are many possibilities; in the expert's mind there are few." The admission of uncertainty—of not having a full grasp of a given subject or situation—is understood as a position of intellectual and emotional freedom, as it suggests capaciousness rather than close-mindedness. The large-scale imaginings of ecological enmeshment often ask their learners to dwell in uncertainty, and the beginner's mind allows students to entertain the unstable terrain of sustainability with flexibility and receptivity. As Iyengar, one of the twentieth century's most important teachers of yoga, writes, "You are a beginner in yoga. I too am a beginner from where I left my practice yesterday. . . . I don't want yesterday's experience. I want to see what new understanding may come in addition to what I had felt up to now."[7] The beginner's mind mediates an experiential identification between student and teacher, and thus it is necessarily inclusive, as it allows individuals to access the practice from any position. The beginner's mind is reinforced by a commitment to daily practice, which facilitates diurnal awareness, or awareness of the subtle changes that each new day presents. Daily practice requires discipline and commitment, and as such it is a useful concept that encourages students to establish their own long-term practices and identities.

Both the beginner's mind and daily practice concern helping students feel comfortable accepting their own positions. Importantly, the instructor must also level herself and identify as a co-participant in this process. More and more, I find my yoga practice encourages me to open my heart in my classroom, as I speak more freely of my process and path. Sharing at this level helps to comfort students who may feel uncertain, isolated, or confused by the many unknowns surrounding them. It is reassuring to them to understand that the meaningful value is in process and not to be overly concerned with product, the value of which is fleeting at best. Like coming into a familiar yoga pose, the humanities classroom offers an invaluable opportunity to return, to accept, to reveal, and to receive. Each session offers a variant dynamic and situational context, and the beginner's mind helps draw attention to the present moment of engagement. Consider revisionary workshopping, for example: revisiting work from the perspective of the present moment reveals how the changes within us project upon all that we see and do.

While much attention in sustainability has been focused on the sciences, the humanities offer a highly valuable space for creating experimental ecological collectives where students can reflect and meditate on their own attitudes, judgments, and choices. In *The Ecological Thought* (2012), Timothy Morton argues for the importance of meditative practices in ecological collectives:

> Ecological collectives must make space for introversion and reflection, including meditative practices. . . . If we take seriously the charge that the problem with science isn't the ideas it develops but the attitudes it sustains, then ecological society must work directly on attitudes. This means, ultimately, working on reflection, and this means meditation, if it's not just to involve replacing one set of objectified factoids with another.

Morton goes on to explain, "Meditation implies an erotics of coexistence, not just letting things be. Meditation is yoga, which means yoking: enacting or experiencing an intrinsic interconnectedness."[8] And with the experience of "intrinsic interconnectedness" comes the accordant awareness of personal responsibility.

Systems thinking is another way of examining this interconnectedness. In her book *Thinking in Systems: A Primer* (2008), Donella Meadows points out that within systems design and analysis is the need for sustainability, particularly consideration of the ways in which people live in the systems. As she writes, "Living successfully in a world of systems requires more of us than our ability to calculate. It requires our full humanity—our rationality, our ability to sort out truth from falsehood, our intuition, our compassion, our vision, and our morality." One key piece of "systems wisdom" that she identifies concerns locating "responsibility within the system":

> Ever since the Industrial Revolution, Western society has benefited from science, logic, and reductionism over intuition and holism. Psychologically and politically

> we would much rather assume that the cause of a problem is "out there," rather than "in here." It's almost irresistible to blame something or someone else, to shift responsibility away from ourselves, and to look for the control knob, the product, the pill, the technical fix that will make a problem go away.[9]

Consciousness and acceptance of personal responsibility is arguably the most critical problem in sustainability education—and it directly touches the lives of undergraduates in so many ways, not the least of which is the issue of consumption. Today's students live in a world of ever-changing iPhones, tablets, laptops, etc., and are quite comfortable communicating through these devices. Often this tendency toward mediated communication manifests in discomfort with traditional interpersonal relationships, including classroom discussions. More than any other generation in history, ours is influenced by the products of industrial capitalism, relentless marketing, and media frenzy. In turn, students exhibit a self-consciousness regarding this influence, and when given the opportunity, they are often ready to talk passionately about their material consumption choices—including food, clothes, cars, and techno-gadgetry. As young people shape their identities, they confront the ways in which those identities are marketed and purchased, as well as the ways in which creativity and mindfulness might allow them freedom to negotiate away from financial strictures. In other words, by locating and accepting responsibility for the ways in which their behaviors are influenced, students may empower themselves to transcend those influences.

In addition to consumption is the issue of competition over cooperation. Large research universities, and mainstream education in general, mandate competition among students. Whether for grades, jobs, or entrance into graduate school, the professional pressures of the corporate academy increasingly focus student attention on accumulating high grades and test scores at the expense of any sense of cooperative spirit. This combination of consumption and competition are two forces that root their ways into the minds and bodies of a student population. And, in my experience, they can lead to instability and unhappiness among young people. Indeed, Stony Brook University was featured in a 2010 article in the *New York Times* on declining mental health among college students. As the article explains, "Forty-six percent of college students said they felt 'things were hopeless' at least once in the previous 12 months, and nearly a third had been so depressed that it was difficult to function, according to a 2009 survey by the American College Health Association."[10]

Sadly, I have witnessed vigils for more than one student's suicide, and just about every semester during office hours, I find myself shepherding students to the counseling center after hearing their admissions of anxiety, panic, and despair. They often come to me to discuss these issues because my courses open up conversations that yoke intellect and emotion. In response to a student body that seems increasingly self-conscious of its own anxiety, I have begun to integrate more meditation practices into the classroom and to advocate openly for self-care as a necessary part of academic success. I often take a few moments to talk about the importance of drinking water, eating well, and getting adequate sleep—and reinforce to my students that the quality of their work is intrinsically connected to their physical and mental health.

In terms of sustainability, it seems unrealistic to ask students to take risks in changing their consumption patterns and competitive tendencies when they are often sick, stressed, emotionally paralyzed, and struggling to find ground to stand on. Yoga and mindfulness practices draw attention to these states, while also offering widely accessible tools that can help students to find relief from these conditions. For example, I recently started a class session by instructing students in a simple meditation that emphasizes posture, in which the students sit upright at their desks, close their eyes, and breathe deliberately. After the brief interlude, one student observed that he had not realized how difficult it was to sit up straight; he had no awareness of his own physical posture and had never been asked to draw attention to it as a college student. Similarly, many students admit to having no awareness of their breath. This endemic disconnect between body and mind makes navigating the education system unnecessarily difficult—and it creates an almost insurmountable obstacle on the path to becoming healthy human beings in a healthy world.

Students often respond with enthusiasm and interest when asked to practice a form of meditation, and I have found that many students are drawn to yoga and meditation because they seek change in their personal lives. Their points of entry are varied, whether to deal with stress or anxiety—or sometimes because of a desire to become more flexible, strong, and/or physically attractive. Whatever reasons draw them to yoga, the practice is adaptable and eventually asks them to reflect upon the reasons for their initial interest.

Possibly the most powerful tool that yoga and mindfulness offer in creating a more sustainable higher-education environment is the idea of *praxis*. The humanities seem to be uniquely suited for manifesting this endeavor, as the skills being developed are critical thinking and writing. In a well-managed workshop environment, all students are empowered to have voices and make contributions. In turn, I often share my own creative work with students in order to show them that I too am in process as a writer. This form of integration may be thought of as a form of engaged pedagogy. As bell hooks writes, "Engaged pedagogy emphasizes mutual participation because it is the movement of ideas, exchanged by anyone, that forges a meaningful working relationship between everyone in the classroom." It is a practice that "establishes a mutual relationship between teacher and students that nurtures the growth of both parties, creating an atmosphere of trust and commitment that is always present when genuine learning happens."[11]

When introducing students to the problems and possibilities surrounding sustainability, instructors must be conscious of the ways in which they practice what they preach. Returning to Hanh's words that open this chapter, there is a critical difference between perpetuating a culture of blame and creating a culture that is engaged. For these reasons, instructors must carefully question the values that inform their classroom practice. As Toni Morrison asserted in a powerful speech she offered at Princeton University, "It becomes incumbent upon us as citizen/scholars in the university to accept the consequences of our own value-redolent roles. Like it or not, we are paradigms of our own values, advertisements of our own ethics—especially noticeable when we presume to foster ethics-free, value-lite education."[12] In teaching sustainability and ecological thinking, the importance of these paradigms seems even more heightened and explicit. In guiding students to manifest change, instructors must be personally willing to do this work, thus creating a collective of co-participants and open dialogue.

But what does this all look like in practice? Primarily, the pedagogical practice begins with the construction of a syllabus and the creation of dialogue within the classroom. In the following section of this essay, I will discuss four courses that I have developed, all of which encourage students to integrate mindfulness practices into their intellectual development while also fostering dialogue concerning contemporary sustainability issues.

Integrating Mindfulness Practices into Sustainability Pedagogy

I have designed and taught a foundational course titled "Interpretation and Critical Analysis" for students majoring in environmental humanities in a formal sustainability studies program. We start the semester by reading *Walden* followed by key literary texts that emphasize sustainable practice. Guided by Thoreau's metaphorical resonance in "The Bean Field," the course focuses on gardening and sustainable agricultural practices, as outlined by writers like Wendell Berry and Michael Pollan. Both Pollan and Berry speak from the position of practitioners for whom the garden is an adaptable metaphor that integrates the ethics of care with both practical and theoretical knowledge. In this course, students keep a handwritten journal that can be about any subject, so long as it is kept, ideally, on a daily basis. Building on mindfulness established through journaling, their final project asks them to practice a theoretical idea or argument that they have developed. Students have installed gardens, attempted to grow plants in their dorm rooms, and coordinated the planting of trees on campus, among other activities. The larger pedagogical imperative in this course encourages students to be conscious of the metaphors they choose to organize their language and experience. Through the example of the garden, students are able to analyze and reflect upon a versatile metaphor that speaks to different disciplines, while the experiential component galvanizes theory into reality.

"Eating Mindfully" is a course I designed and teach as both a freshman seminar and (in a slightly expanded form) an honors course that serves as an introduction to food studies framed directly through the standpoint of mindfulness. Usually, students' initial expectations for the course involve nutrition and what food can do for them. The intervention of mindfulness asks them to consider how their food choices not only affect their bodies, but also the health of their environments and communities. They are required to keep a food journal throughout the semester. Their first project asks them to document the food pathways of a favorite meal, considering season, miles traveled, packaging, production, etc., as well as the limitations and challenges surrounding such questioning. Because food is a subject that is deeply personal, I integrate a creative component in which students write either a brief food memoir or a poem that celebrates their cultural food identities. The goal in this course is not to convert the students to a particular ideology, but rather to encourage them to diversify the ways in which they consider their own choices. Because psychological attachments and addictions often belie conscious food choices, I want students to be comfortable examining their own attachments without fear of judgment or shame. Thus the food memoir assignment allows them to celebrate their personal background, while the food pathways project asks them to interrogate analytically something they tend to take for granted. We also spend time reading Kevin Young's beautiful anthology of poetry, *The Hungry Ear* (2012), to generate further attention to the lyric experience of food. The sensory attention of poetic language helps widen awareness and appreciation for the experience of eating. Another text for this course is Hanh's thin volume *How to Eat* (2014), in which he offers a series of short meditations on "Eating in Silence," "Cooking without Rushing," and "Choosing What to Eat," among others. Each session we read one meditation, and I ask the students to attempt to practice it at some point in the near future, recording their observations in their journals.

Each time I teach "Eating Mindfully," I encounter resistance on the part of students to accept the personal responsibility that comes with food choices. The course is underpinned by the Buddhist virtue ethic that frames mindfulness practices, and foundational to sustainability is understanding that personal choices do make a difference. As Hanh writes, "The planet suffers deeply because of the way many of us eat now. Forests are razed to grow grain to feed livestock, and the way animals are raised pollutes our water and air. A lot of grain and water is also used to make alcohol. Tens of thousands of children die of starvation and malnutrition every day, even though our Earth has the ability to feed us all." He continues, "With each meal, we make choices that help or harm the planet. 'What shall I eat today?' is a very deep question. You might want to ask yourself that question every morning."[13] Students observe that although Thich Nhat Hanh's writings seem very simple, enacting his suggestions is quite challenging. They then must confront the question of *why* these simple ideas are so challenging—exploration that opens the door for analysis of how society can influence decisions and how mainstream comfort zones such as television and social media may create obstacles to self-reflective learning.

Another course in the sustainability studies curriculum in which I integrate mindfulness practices is titled "Extreme Events in Literature," and here the very nature of the material—natural and human-initiated disasters, as well as contemporary instances of terrorism—is traumatic and emotionally charged. Given such difficult subject matter, the course emphasizes the work of healing and transformation, following Rebecca Solnit's sense that "if paradise now arises in hell, it's because in the suspension of the usual order and the failure of most systems, we are free to live and act another way." Here Solnit suggests that in situations of disaster, compassion and cooperation supplant the indifference and competition of the "usual order."[14] Analyzing the rhetorical and representational strategies of extreme events, the course brings a critical eye to bear on doomsday prophesizing and rhetoric that incites fear, terror, and panic; our examination of this topic focuses not just on the events themselves, but also on the community building and renewed ethical relations that emerge from them. Our study is guided by key questions: What opportunities do such events afford for transformation and renewal? What can environmental humanists learn from studying such events? And how can we, as writers, thinkers, and community members, take personal responsibility for our languages of response to, and methods of engagement with, such extremity?

The primary text in "Extreme Events" is Karen Lofthus Carrington and Susan Griffin's *Transforming Terror: Remembering the Soul of the World* (2011), and as the semester progresses, class discussions turn more and more toward the work of healing and mediating trauma. "Extreme Events" ultimately becomes about reconciliation, compassion, cooperation, forgiveness, and nonviolence. As Griffin and Carrington write, "Since essential resources such as water are affected by climate change, and environmental crisis also threatens to become a major cause of violence and warfare, sustainability and peace are inseparable."[15] The first time I taught this course, I was fortunate to have a small and very engaged group of students, and following their request, we dedicated one class session to a full yoga practice in the classroom. On other occasions, we practiced nonviolent communication, and while our classroom discussions were guided by texts, they were also rooted in the daily observations of the students as we analyzed their interpretations and responses to events such as Hurricane Sandy. In another professor's course on disasters, the students learned how to assemble a survival pack, which may be relevant to a military unit or even the Boy Scouts, but within the context of a sustainability studies classroom, it seems somewhat reactionary. In my course, we thought about the importance of patience, clear communication, leadership, breathwork, and other meditation practices that might allow students to be more effective members of their communities amidst a crisis. They exercised how to respond in a morally responsible way.

Mindfulness practices and sustainability theory can be integrated into just about any course. For example, "World Literature: Ancient to Early Modern" is a general-education elective that fulfills a humanities and global thinking requirement. This course, on the surface, has little to do with sustainability. However, through carefully selected texts that frame individuals as moral agents within a larger cosmic framework, the course opens up questions and dialogues concerning environmental stewardship, personal responsibility, membership in a biotic community, consumption patterns, intergenerational justice, and virtue ethics. I have found that the ancient texts foster compelling discussions that encourage students to develop informed responses to changes in both natural and societal phenomena induced by climate change and current issues of sustainability. The classics reveal that despite the particular urgency and immediacy of climate change, human beings have long theorized and negotiated the ethical problems concerning relationships between people, places, and future generations. Didactic poetry, in particular, which self-consciously attempts to create continuity and community through narrative, proverbial wisdom, metaphor, and poetic imagery, presents a rich and accessible history framing these problems and responses.

Ancient texts are paired with contemporary writing in this course; for example, Virgil's *Georgics* are read alongside Wendell Berry's writing on sustainable agriculture, *The Bhagavad Gita* exists alongside excerpts from Thoreau, Ursula Le Guin's translation of *The Tao Te Ching* joins a selection of her own nonfiction, and W. S. Merwin's poetry partners with his translation of Muso Soseki's poetry. The pairings allow students to see how alive and important the ancient texts are to contemporary literary activism. Also, in each unit, we engage in different forms of meditation and creative writing. The students are asked to respond freely and openly to all exercises and to constantly "try on" the precepts that the texts posit. How do they interpret the words on the page? What would it look like to follow the *Gita*'s call for renunciation? In today's ultracompetitive world, how would one implement the *Tao Te Ching*'s assertion that by "not competing," wise souls "have in all the world no competitor"?[16] When we study a selection of Horace's *Odes*, students are briefly introduced to the structure of the Epicurean Academy, where value was placed on working toward *ataraxia* or spiritual tranquility, and virtues like friendship, gratitude, and patience underpinned the educational ethos. We have lively conversations in the classroom about the value of such virtues—and their relative absence in contemporary education. Out of all the classes I have taught that add mindfulness into the fold of learning, this course seems the most successful, perhaps because of the ways in which the classics rhetorically frame the urgency of moral agency and personal responsibility. It is also empowering for the students to recognize the ways in which they can directly connect to ancient texts and use them as guides to frame experimental forays into meditation.

Finding Peace within the Frenzy

In *Earth in Mind: On Education, Environment, and the Human Prospect* (2004), Orr writes:

> The plain fact is that the planet does not need more successful people. But it does desperately need more peacemakers, healers, restorers, storytellers, and lovers of every kind. It needs people who live well in their places. It needs people of moral courage willing to join the fight to make the world habitable and humane. And these qualities have little to do with success as we have defined it.[17]

These lines have been popularly reproduced and often quoted (and misquoted), perhaps because they speak so directly to a particular reality of contemporary education that prioritizes economic success above all else. The practices of yoga and meditation encourage people to live well in their places while bolstering their moral courage (not to mention their emotional and physical health). The humanities add an integral dimension to sustainability education because they can integrate reflective and meditative practices to better understand ecological connection and the personal responsibility that comes with ecological collectivity. Important, though, are the many steps that individuals and organizations are making to further these discussions. The Center for Contemplative Mind in Society publishes a peer-reviewed journal and organizes retreats and conferences; the Omega Institute in Reinbeck, New York, hosts an annual program on mindfulness in higher education; and academic programs such as Creativity & Consciousness Studies at the University of Michigan work to bring mindfulness into curricula. More and more, the value of contemplative studies is recognized and integrated.

Although I was advised to mask associations with yoga to navigate the job market, to do so would alienate me from my greatest source of strength as an intellectual, an educator, and a human being. And while the value of yoga and meditation are increasingly recognized by the medical community for both their preventative and therapeutic applications, their extended integration in more mainstream academic classrooms holds great hope for the future, particularly as a means of helping to integrate the interdisciplinary fragmentation of sustainability studies. The urgency of sustainability, paradoxically, requires a capacity to slow down: to slow down consumption, to deeply contemplate before acting, and even more importantly, to accept personal responsibility for our actions, at this moment, right now. The practice of yoga aligns with these intentions, as stated by Patañjali's first two yoga sutras. The first sutra offers hope and direction for modern challenges such as climate change, while framing the experience in the present tense: *atha yoganusanam* ("Now, the teachings of yoga"). Right here, right now, we have the opportunity (and responsibility) to begin. And what is it that we are embarking upon, the second sutra answers: *yogas citta-vrtti-nirodhah* ("Yoga is to still the patterning of consciousness").[18] The modern brain is embattled with a persistent, seemingly endless barrage of stimuli, some relevant, most merely distractions. Yoga works to develop the skill set to manage this phenomenon, to find peace within the frenzy.

Currently, higher education relies on overworked, underpaid, and contingent adjuncts to teach exhausted, overstimulated, and stressed-out students to reach for economic success in an ultracompetitive, industrial capitalist marketplace. The inherent strains of this system are contributing to its irrelevancy to current issues, as well as its disintegration from within. Integration of yoga, meditation, and mindfulness practices, as well as necessarily redirecting the focus of the system towards a more balanced worldview, incorporating cooperation, empathy, and an understanding of the human condition, may go a long way towards revitalizing an invaluable higher education system that could play a crucial role in guiding humanity through the challenges inflicted by the past two hundred years of "progress." Climate change mandates that current trends in both consumption and production must change, and a sustainable education system that integrates yogic practice and mindfulness is a critically important place to start.

NOTES

1. Margaret Robertson, *Sustainability Principles and Practice* (New York: Routledge, 2014), 3.
2. David W. Orr, *Earth in Mind: On Education, Environment, and the Human Prospect* (Washington, DC: Island Press, 2004), 8.
3. Henry David Thoreau, *Letters to a Spiritual Seeker*, ed. Bradley Dean (New York: W.W. Norton, 2005), 50; Thoreau, *Walden*, ed. Jeffrey S. Cramer (New Haven, CT: Yale University Press, 2004), 322.
4. Stanley Cavell, *The Senses of Walden* (Chicago: University of Chicago Press, 1992), 67.
5. E. B. White, "A Slight Sound at Evening," in *Nature Writing: The Tradition in English*, ed. John Elder and Robert Finch (New York: W.W. Norton, 2002), 441; Thoreau, *Walden*, 42.
6. Gary Snyder, *A Place in Space: Aesthetics, Ethics, and Watersheds* (New York: Counterpoint, 2008), 49.

7. Shunryu Suzuki, *Zen Mind, Beginner's Mind* (Boston: Shambhala Press, 1987), 2; B. K. S. Iyengar, *The Tree of Yoga* (Boston: Shambhala Press, 2002), 73.
8. Timothy Morton, *The Ecological Thought* (Cambridge, MA: Harvard University Press, 2012), 127.
9. Donella Meadows, *Thinking in Systems: A Primer* (White River Junction, VT: Chelsea Green Publishing, 2008), 170, 179, 4.
10. Trip Gabriel, "Mental Health Needs Seen Growing at College," *New York Times*, December 19, 2010.
11. bell hooks, *Teaching Critical Thinking: Practical Wisdom* (New York: Routledge, 2013), 21, 22.
12. Toni Morrison, "How Can Values Be Taught in the University?," *Michigan Quarterly Review* 40, no. 2 (Spring 2001): 6.
13. Thich Nhat Hanh, *How to Eat* (Berkeley, CA: Parallax Press, 2014), 58–59.
14. Rebecca Solnit, *A Paradise Built in Hell: The Extraordinary Communities That Arise out of Disaster* (New York: Viking, 2009), 7.
15. Karin Lofthus Carrington and Susan Griffin, *Transforming Terror: Remembering the Soul of the World* (Berkeley: University of California Press, 2011), 253.
16. Lao Tzu, *Tao Te Ching: A Book about the Way and the Power of the Way*, trans. Ursula Le Guin (Boston: Shambhala Press, 1997), 31.
17. Orr, *Earth in Mind*, 12.
18. Chip Hartranft, trans., *The Yoga-Sutra of Patañjali: A New Translation with Commentary* (Boston: Shambhala Press, 2003), 2.

BIBLIOGRAPHY

Carrington, Karin Lofthus, and Susan Griffin. *Transforming Terror: Remembering the Soul of the World*. Berkeley: University of California Press, 2011.

Cavell, Stanley. *The Senses of Walden*. Chicago: University of Chicago Press, 1992.

Hanh, Thich Nhat. "The Bells of Mindfulness." In *Moral Ground: Ethical Action for a Planet in Peril*, edited by Kathleen Dean Moore and Michael P. Nelson, 79–81. San Antonio, TX: Trinity University Press, 2010.

———. *How to Eat*. Berkeley, CA: Parallex Press, 2014.

Hartranft, Chip, trans. *The Yoga-Sutra of Patañjali: A New Translation with Commentary*. Boston: Shambhala Press, 2003.

hooks, bell. *Teaching Critical Thinking: Practical Wisdom*. New York: Routledge, 2013.

Iyengar, B. K. S. *The Tree of Yoga*. Boston: Shambhala Press, 2002.

Lao Tzu. *Tao Te Ching: A Book about the Way and the Power of the Way*. Translated by Ursula Le Guin. Boston: Shambhala Press, 1997.

Meadows, Donella. *Thinking in Systems: A Primer*. White River Junction, VT: Chelsea Green Publishing, 2008.

Morrison, Toni. "How Can Values Be Taught in the University?" *Michigan Quarterly Review* 40, no. 2 (Spring 2001): 273–278.

Morton, Timothy. *The Ecological Thought*. Cambridge, MA: Harvard University Press, 2012.

Orr, David W. *Earth in Mind: On Education, Environment, and the Human Prospect*. Washington, DC: Island Press, 2004.

Robinson, Margaret. *Sustainability Principles and Practice*. New York: Routledge, 2014.

Snyder, Gary. *A Place in Space: Aesthetics, Ethics, and Watersheds*. New York: Counterpoint, 2008.

Solnit, Rebecca. *A Paradise Built in Hell: The Extraordinary Communities That Arise out of Disaster*. New York: Viking, 2009.

Suzuki, Shunryu. *Zen Mind, Beginner's Mind*. Boston: Shambhala Press, 1987.

Thoreau, Henry David. *Letters to a Spiritual Seeker*. Edited by Bradley Dean. New York: W.W. Norton, 2005.

———. *Walden*. Edited by Jeffrey S. Cramer. New Haven, CT: Yale University Press, 2004.

White, E. B. "A Slight Sound at Evening." In *Nature Writing: The Tradition in English*, edited by John Elder and Robert Finch, 440–448. New York: W.W. Norton, 2002.

Ecological Journeys: From Higher Education to the Old Farm Trail

Barbara George

> He had resented injustice and cried out against that sin of sins, the degradation of man by man, believing the world held few things more precious than human dignity.
>
> —Thomas Bell, *Out of This Furnace*

I had driven up and down one mountain and was now reoriented, following the steep curve away from Bridgeville, a small borough outside Pittsburgh. Mill houses lined the hill to my left, and to my right, separated by a flimsy guardrail, was a wooded ravine below, gray on this bitingly cold January day. I gripped the steering wheel and looked over the edge, spotting white sycamores and rusting equipment among the winter trees.

Orientation

A few miles away the land flattened, and newer, spacious suburban homes dotted the landscape. I turned into the Boyce Mayview Park. There could be no doubt that this was my destination: the bottom of the hill was marked with a new and impressive stone sign, and crowning the hill arose a brand-new glass and brick community center. Behind the center lay well-groomed softball and soccer fields—quiet now, but poised for action with the return of warm weather.

Inside, through the lobby's steamed-up glass walls, I could see residents of all ages taking advantage of the indoor pool. It seemed a moment of respite, of human triumph over the elements, but it was here that I was redirected outdoors in my quest for the nature center, with the instructions "Go back outside, and follow the signs to the trailer." Trailer? I was attempting to find a position to augment my current English adjunct position. A reading program through which I had been teaching several courses had been abruptly dropped, and thus my colleagues and I suddenly lost several courses we were relying on for the spring. I had seen an ad for a part-time environmental educator, and I jumped at the chance. Trailer or no, I needed to work.

Accordingly, I drove away from the gleaming center, following a series of smaller signs that directed me to the Outdoor Classroom's office, situated in two old trailers and located a distance behind the larger Boyce Mayview Park. The community center, I later learned, served the Upper St. Clair area. "Upper" is the operative term: Pittsburgh's Upper St. Clair Township is known for good schools, an exploding housing market, and, more recently, attention to the local park system. I would later learn that considerable resources had been put into rehabilitating this land, with both an agricultural and industrial past, into a place that would provide recreation opportunities for the township.

As I had viewed the map prior to my first drive in, names like "Coal Run" and "Coal Pit Run Road" jumped out at me. Artifacts from previous ways of interacting with the land were revealed in both the names and the landscape itself: farmhouses and fields, mill houses, coal piles, and train tracks mixed with newer suburban developments and shopping plazas. On that January day, I physically navigated my way to my interview by passing through these worlds. But it was my later work at the Outdoor Classroom that led me to consider time and space, and various ecologies situated within them, more clearly.

Reinhabitation

I made this journey to the Outdoor Classroom because my experience with the university system as an adjunct instructor was not sustainable in a very simple way: the job did not support me financially. On my English adjunct teaching days

I drove west into Ohio. Today, I drove from Ohio into Pennsylvania. As I passed the reminders of a past industry on my way to the interview, I mused about the workers' movements through time that had occurred only miles from the Outdoor Classroom. The irony of seeking work near Pittsburgh with its significant labor movement history (the Homestead Strike in 1892 and the McKees Rock Strike in 1909) was not lost upon me. Nor was the more recent way in which Pittsburgh had once again been thrust into the spotlight in terms of labor. Daniel Kovalik's "Death of an Adjunct," detailing Margaret Mary Vojtko's experiences as an adjunct at Duquesne University, had been published in the *Pittsburgh Post-Gazette* in 2013.[1] The piece sparked a national outrage and debate about the plight of adjuncts: the lack of security and benefits, and the low pay for contingent labor that was increasingly the backbone of higher education. The account of Vojtko's experiences, I knew, had been contested since the piece had been published, but the reality of lack of security was still all too real as I made my way to this interview.

I hoped the opportunity to teach at the Outdoor Classroom would augment financially my teaching of English composition. But I also anticipated that an experience with the Outdoor Classroom would more profoundly question the decontextualized "management" of literacies that I had unwittingly become a part of at the university. David Orr suggests that "ecological literacy, further, implies a broad understanding of how people and societies relate to each other and to natural systems."[2] This assessment of literacy intrigued me. I wanted to be part of a longer-term community, rather than drift as a transient part of a learning institution that severed our relationship after every eight or fifteen weeks or so. Moving from institutional buildings and rooms at the university, with little opportunity to interact with other instructors, I missed identifying with a particular "place" and wanted to intimately connect with a landscape. The adjunct experience led me to question, more deeply, just what is it that we sustain when we teach.

Peter Goggin, in exploring rhetorics of sustainability, investigates many scholars' definitions of sustainability and offers the following from Derek Owens: "Hastily defined, sustainability means meeting today's needs without jeopardizing the well-being of future generations." The term, however, is slippery. Orr, in addressing "ecological literacies," suggests "the word 'sustainable' conceals as much as it reveals." Orr goes on to explain that "hidden beneath the rhetoric are assumptions about growth, technology, democracy, public participation, and human values." Similarly, John Dryzek points out that it is only by explicitly addressing those assumptions that this term "sustainability" moves beyond a buzzword or slick greenwashing.

Dryzek outlines stakeholders within the "pillars" of sustainability—economic, social, and environmental—that might co-opt sustainable development into a system that maintains the status quo, in which already powerful global players define sustainable development "in terms favorable to them." While Goggin cautions about the limits of analyzing sustainability without context or notion of praxis, he ultimately suggests critical investigation of sustainability and urges in the analysis a redefinition of the "literate self" in ways that "sustain the humanistic ideal of critical reflection."[3]

I was reminded, often, of these notions of critical reflections about sustainability in the months that followed my first visit to Boyce Mayview Park. I landed the job at the Outdoor Classroom and, along with three other new employees, began working as a "casual facilitator" in the spring. This designation meant I did not have regular hours, but I could sign up for openings each month, which ironically suited my university teaching schedule perfectly: my loss of adjunct work spurred the taking on of additional adjunct work.

Early on, I was able to recognize place in ways from which I had often become disconnected while at the university. We acknowledged, in our Outdoor Classroom lessons, how land "spoke" through the ways material and space were reimagined through space and time. One trail, "Between Two Worlds," followed a ridge that had once been a rail line, built to haul coal. As we walked the trail during a training session, remnants of an industrial past—partially rusted metal engines whose functions we could only guess—emerged in the woods. Below the ravine was Chartiers Creek, where the sycamores grew, their tips reaching up to trail level. In early spring, we saw the massive nests of a heron rookery below us as we walked along the old coal-cart trail. In the summer we would see herons in flight above us, scouting for food in Deer Meadow by the trail. The place was dynamic. Contradictions abounded: the junkyard of sorts abutted the largest heron rookery in western Pennsylvania. Spatial scales constantly shifted through various time scales. In contrast to some of the content or literacy practices I taught at the university, the Outdoor Classroom lessons did not shy away from what was apparent—the remnants of the past landscapes existing in current landscapes—but also lives that once inhabited, and still do inhabit, that space.

These immersions in contexts of place allowed us to discuss how places shaped us and how we shaped places, particularly in southwestern Pennsylvania. It was obvious from my drive in on the first day that interspersed between the new housing developments were traces of earlier heavy industry; after all, this was the Rust Belt.

In an ironic transformation, this newly affluent community had the resources to designate a place to study the outdoors. What was not as immediately visible was that the land was part of a less equitable past haunted by sites of environmental injustices that Rob Nixon calls a "slow violence."[4] The Outdoor Classroom encouraged us to weave these industrial artifacts into our lessons, and I appreciated that the lessons allowed instructors to explore these past orientations. These lessons helped me to reflect upon the research-writing and business-writing courses that I had taught in Ohio as an adjunct. Both Ohio and southwestern Pennsylvania were places that experienced boom-and-bust cycles of industry. Without the critical reflection, I wondered if I might, in my writing classes, be perpetuating systems that did not account for risk assessment over longer periods of time and for various species. And even if I did begin conversations about these subjects in those classes, to what effect could students "speak" in systems that were situated toward new technologies that were emerging in both Pennsylvania and Ohio, like fracking, or continued older ones, like coal, that had long-term consequences on human and ecological health? Like the unsustainable system of adjunct labor, these industrial activities also were often tacitly, or even overtly, legitimized by universities. At the Outdoor Classroom the long and deep past of an industrial system was a contextual element that was ever present. The traces of past industry perhaps were "hiding" in a well-to-do neighborhood or on the bucolic grounds of the Outdoor Classroom, but those traces were there.

Other experiences of past industries were more subtle. On a community weekend, we wrestled an ancient toilet and bathtub from the underbrush behind the Learning Lab to new native gardens by the office, laughingly carrying them across Hawk Meadow in a team of six. Finally, toilet and tub stood in the garden transformed as planters rather than remaining junk that had been discarded. This event led me to think about by what forces we might resituate the deep contexts that we had inherited. Through an ecology of writing, Marilyn Cooper focuses on the dynamism of ecology in terms of time, space, and the socially constructed creation of environments. She describes composition by discussing ways that humans create and are created by their environments. Rachel Tillman expands this concept of ever-changing ecology with the notion of active materialisms that highlight the interactions between cultural forces, materials, and the environment.[5] Illustrating these theories, the material land and our experiences with it were constantly being negotiated in the Outdoor Classroom spaces. Other remnants, besides the toilet and the bathtub, of an agricultural past also appeared in the woods framing the fields: multiflora rose shrubs, once planted as a living fence barrier by past farmers, had taken on their own agency and now grew throughout the woods. We now labeled them an invasive, but the shrubs were a part of place. Barbed wire, too, appeared—sometimes in woods that had once been field, but sometimes growing as part of a tree itself—pointing to complicated relationships between culture and nature that were not so easily "cleaved."

Perhaps most telling of the human lives that once inhabited the space are the names of the Old Farm Road Trail and Lost Farm Trail. Both trail names refer to the farm that was once part of the Mayview State Hospital, located on the site since the nineteenth century. At its height, Mayview (once called Marshlea, after the London poorhouse that housed Charles Dickens's father, an allusion to Mayview's history as a "poorhouse") served more than four thousand patients in an imposing brick building and numerous outbuildings. Patients worked at various duties on the farm, including a working coal mine. While definitions of sustainability vary, in terms of food resources, Mayview was remarkably sustainable: for example, it once produced more than 60 percent of its own food.[6] The classroom trailers stood at the edge of many fields that had once produced food, with only the trail names a reminder of what had been. A few black-and-white pictures on the Outdoor Classroom's office walls showed some farm outbuildings. Mayview closed in 2008, and several outbuildings were demolished more recently. New models of mental health enabled a move away from institutions toward home- and community-based services. The fields that once fed southwestern Pennsylvania's poorest and most vulnerable had been graded and reseeded, reappearing as soccer fields for children of the wealthy. But ghosts of the people in the past remained in pictures on the Outdoor Classroom's wall and on the land. Like artifacts found in a language that suggests certain social arrangements, we constantly found material artifacts of place that suggested a certain orientation toward the land: a rusted barbed-wire fence in the woods, a fragment of a broken dish from Mayview in the stream.

As instructors, we never tried to hide what this land had once been. And in our introductions to the Outdoor Classroom on the steps of the Learning Lab, we addressed the history of the place, including acknowledgment of the Mayview State Hospital. Still, the comments from students could be unsettling. "It was a crazy hospital!" a nine-year-old boy yelled during one introduction. An instructor diplomatically changed the conversation. The student's words revealed the stigma that still exists about those in society who do not fit into norms and who must be "managed." The farm refuse that appeared showed this land to have been a place to dispose of what was undesirable. This disposal became a metaphor for what is valued socially, and is tied to how humans are valued. In the same way, I was haunted by some of the old pictures I had seen of Mayview Hospital patients. The few pictures offered some glimpses of the realities of living in such an institution. In one, three women looked out, gazing vacantly above a huge pile of green beans that had been shelled. Two children, heads shaved, smiled innocently. A little black boy and a little white boy stood side by side. *What was life like for those women? How had they come to be at Mayview? What were those children's stories?* Unwed mothers, I read, had once been sent to Mayview, as they had been seen as an "abnormality" or "mother insane."[7] Could that have been the story of these women shelling beans? And what was the story about race at this place and time? While I could not compare my institutionalized experience of an adjunct directly to these people's experiences, perhaps my own experience deepened my sympathy for them. What is it like to be a contributing part of a system that tacitly does not value you, I asked, or values you only as a kind of "resource"?

As a reading and composition instructor, I taught "literacy" to students who came from and would travel through several different disciplines and larger communities: personal, civic, and academic. While I hoped that the literacy practices I taught might enable students to question critically their institutions and institutional systems, my position as an adjunct made me a "sponsor" of literacy myself. This tenuous position proved, ironically, to be ultimately unsustainable. We simply weren't valued. Not only because I could not teach the number of classes needed to pay my bills, but because instructors in university learning communities struggled to find ways to engage in research that would make more equitable the very systems in which we, as adjuncts, found ourselves. Changing higher education in favor of more equitable learning experiences for students was also problematic. I might facilitate the opportunity for students to write about rising student-loan debt in classes, but how could we approach these issues with real institutional praxis in our limited positions? How might this literacy education, like the "land" mentioned by Deborah Brandt, move from a "commodity" to be "managed"?[8] In my studies, I was immersed in the notions of ecologies of literacy that acknowledged contextualization of learning. This immersion led me to explore literacy scholars who were questioning notions of sustainability, and their work led me to ask not only how I was valued, but how students were valued.

From the outset, the Outdoor Classroom offered students the space to "be" on the landscape in ways that traditional classrooms might not capture. We met groups at the Learning Lab, where students sat on the steps that opened to a field and counted bird nests in the joists as we explained the day's itinerary and the few simple rules: Respect wildlife. Respect each other. Don't pass the facilitator on the trail. We often started the day with a couple of bird jokes and ended it with a debriefing of what we'd seen or by simply watching a hawk sail overhead. How different this was from the increasingly onerous syllabus for my university classes, constructed with policies to protect myself against administrative anxiety over student-consumer complaints that had emerged in the university. Surely there was a better way to communicate course rigor and to encourage student inquiry.

Students who came to the Outdoor Classroom were immersed in hands-on experiential learning. The Outdoor Classroom's mission—"Fostering informed stewardship of the environment by highlighting the connection between people and the natural world of southwestern Pennsylvania"[9]—was situated in place-based learning. As I learned the trails and lessons, I began to sense, more deeply, the time scales of the place at the Outdoor Classroom. A few lessons explicitly explored these scales, from the tensions of human inhabitation and animals and plants that lived here (some now gone, and some remaining), to Native Americans who once hunted in the area, to Revolution-era settlers of farms and, later, to farmers and their shallow coal mines, and then to the evolution of strip mines and, now, suburban homes. Through such change, what is sustained in terms of our orientations to the land? I might well ask that question about my own adjuncting experiences.

It is the suggestion of reflection about a sustainable system in conjunction with the work at the Outdoor Classroom that led me to think more critically about my role as an adjunct and a literacy sponsor within a university system. In what ways was my literacy teaching sustainable? Or even resilient? Dryzek shows that "resilient" systems, like sustainability, have many definitions and underlying ideologies that could be interpreted from "business as usual" to "radical approaches to environmental change"—or even, as Dryzek contends, radical economic change no longer dominated by "market liberalism."[10] I felt this distinction of terms keenly as I taught business-writing students to write business reports and plans, knowing my own tenuous position as an adjunct each semester was a result of corporate, economic bottom-line approaches. Indeed, it was a university management approach that did not value the "return" on the university investment that saw the dismantling of a developmental reading program that led me to seek work at the Outdoor Classroom in the first place. While I appreciated critical approaches that might be encouraged in a university classroom, I was also aware of the limitations in assuming that critical discussion was happening, and of the limitations of praxis as a result of that inquiry.

The Outdoor Classroom suggested other orientations towards inquiry. After a year in yet another classroom, or, for some, in organized sports, most students were elated to have a walk on the trail, a day's immersion outside. They were encouraged to touch, feel, and breathe in this space. As an adjunct in an institution, I felt the same way. The students ranged from kindergartners to high school seniors, so on any given day, facilitators might find themselves tying shoes or helping a teen earn an Eagle Scout badge. I was more aware of my students' material selves than I had been in the college classroom, particularly when a kindergartner attempted to run off the trail into a patch of poison ivy, or when a nine-year-old, in the middle of a crayon leaf-rubbing lesson, experienced a violent nosebleed a good mile from the Outdoor Classroom office. (We stanched the flow with cotton balls from the medical kit and the cotton bag that had carried the art supplies, and she happily continued with her leaf rubbings.) We followed lessons, but our ability, as the job description noted, to "'think on our feet' and adapt to teachable moments when they presented themselves" was crucial.[11] We embodied resiliency.

The real beauty of the "classroom" was engagement of the students in the outdoors, from the walk to the lesson site to the lesson site itself, usually a clearing by a meadow, a bridge, or woods, engaging even testy middle-school students in investigating various micro-ecologies. One lesson about observing birds was abandoned when we realized we were being watched by a garter snake, who poked its head out of its hole under a bench to observe us. At other times, minutes slipped away as we dipped nets into what a casual observer might think was a puddle with a few cattails by the parking lot, but a deeper observation revealed as a tiny pond, brimming with life, full of pollywogs, nymphs, bird nests, and traces of deer, surrounded by aquatic plants. There were no textbooks, computer programs, worksheets, benchmarks, assessments, or end-of-course bubble tests. Even the nature guides we handed out at times did not seem to represent the entirety of what we were experiencing. Lessons moved through fluid disciplines; we might write a reflection, draw, compute data, sing, conduct a science experiment, and run on our way or during our lesson—sometimes all of the above. And then, if it rained, evidence of these lessons might turn into a pulpy mess; yet the lessons remained.

In contrast to most adjuncting experiences, the Outdoor Classroom allowed me, as an instructor, to challenge academic distinctions more directly as knowledge-making moved away from constructed notions of atomized disciplines to integrated knowledge. Relationships between ideas and experiences were prominent as edges were blurred and we generated new ideas together. And I was fully empowered to innovate and lead through the lessons. A walk through the Lost Farm Trail that curved around a tiny, narrow valley evoked a Tolkien tale, as students excitedly compared the valley to the movie they had seen. The echoes there were dramatic, and as we sang or yelled to hear them, we might also experience geologic, agricultural, botanic, and biological tales as well.

Ecological theory applied to writing was mirrored by the material experiences I had with students in the Outdoor Classroom. We taught students about diversity in various microsystems and the importance of looking at interactions in the "edges" of the field and the woods. David Barton, a writing scholar, cautions against monocultural views of language and rhetorical situations, encouraging the value at the "edges" of literacy activities. "The edges are its vitality, and variety ensures its future," Barton writes. "Rather than isolating literacy activities from everything else in order to understand them, an ecological approach is needed." The landscape of the Outdoor Classroom and the instructor's and students' approach to it seemed to echo Barton's words: a deep resilience based upon contexts was key. At the Outdoor Classroom, we saw resilience as the land adapted to short- and long-term changes created by interactions between humans and material processes of the land. Again, I recalled Barton's notion of communication ecologies as fluid and changing due to politics and technologies. "New social practices give different possibilities," he asserts, "changing the way [people] communicate."[12]

Excavation

One of the most spectacular walks in the park is a walk on Beech Valley Trail through a ravine down to Chartiers Creek. The beeches, we taught our students, looked like giant elephant legs. We walked with our students, crossing a series of bridges that spanned the steep-sided creek. Through the summer, the ravine sides glowed with wildflowers, notably orange jewelweed or "snapdragons" whose seedpods we taught the children to pop, scattering the seeds. Then, the narrow walls lining the small creek broadened and emptied into the larger Chartiers Creek. We had to leap across this smaller creek to a small, rocky beach. Here, students were encouraged to look for micro- and macro-invertebrates, laughing and splashing and exclaiming over the discovery of a crayfish under a rock.

I wondered, the first time I took a group to this area, why no one had cleaned the beach. It was littered with pieces of brick and of shards of old broken glasses and dishes. Hardly a safe place to undertake the lesson on invertebrates we had been sent down to teach. "Don't touch the broken glass!" I repeatedly cautioned, and I piled up several pieces to dispose of in a more appropriate place. However, every time I returned with another group, a new set of detritus had appeared on the beach. I later learned the entire side of the ravine sloping towards the water had been the hospital dumping ground for over a hundred years. Pieces of old plates, bricks, and farm equipment were continually washed to the beach after a rain, where they rested for a while before eventually being swept into the creek for another journey. The beach's shifting collection made visible what was once "refuse" to be hidden. Very little could be done about it except to incorporate it into a lesson. The "trash" had become part of the ravine, and trees grew over and through it. "Managing" it might involve cutting trees down, which would have devastating erosion effects on the steep hillsides. But these artifacts made this place an example of tensions between culture and nature, and a reminder of humans on this space. I once picked up a curved shard of an old milk-white glass bottle. What did it once hold? For whom? Lotion? Face cream? Medicine? Now it might hold a few grains of sand and a caddisfly larva.

I would have liked to think that past industrial orientations had vanished, and that we had arrived at a more enlightened way of interacting with natural spaces. But we were not able to take students to Chartiers Creek beach for several days one spring because a tanker truck with fracking fluid had toppled off a highway and spilled its contents upstream into Chartiers Creek. Problematic interactions between humans and the land were not a "distant past." We were constantly reminded of these interconnections and interactions, very aware that what happened in one place deeply impacted events in another, that what was often hidden eventually might come to light.

These complex relationships and material ecologies again brought to mind ecologies of language and my own adjuncting experience that were becoming, disturbingly, more common. Of how intertwined language is within social constructs, and, recursively, with constructed materialities that might, at first glance, seem "natural." While Barton encourages new social arrangements based on new communication patterns, James Porter refers to Vincent Leitch's notion of "traces" and "histories" of texts that "[resemble] a Cultural Salvation Army Outlet with unaccountable collections of incompatible ideas, beliefs, and sources." Writing occurs, according to Porter, within constraints of larger social systems or, as he terms them, "discourse communities." Porter elaborates: "We must inevitably borrow the traces, codes, and signs which we inherit and which our discourse community imposes. We are free insofar as we do what we can to encounter and learn new codes, to intertwine codes in new ways, and to expand our semiotic potential."[13] While the Boyce Mayview Park attempted to create a new "space," it was undeniable that histories would appear; the space could not be fully "managed" or "suppressed" to match a cultural construct of what land "ought" to be. Problematic historic and contemporary ways of interacting with land, partially informed by past ideologies, would revisit us. This notion, it seemed, more clearly could apply to my university adjuncting experiences. In my courses we might discuss disparities, for instance, but we would never mention the glaring disparities in the university system itself—not just for adjuncts, but for students increasingly putting themselves in debt for this system.

We might reimagine spaces at the Boyce Mayview Park, but as the Outdoor Classroom made clear, we did so in relationship to what had been in the space before. We might remove buildings, but we couldn't excavate the broken plates and bottle shards that had accumulated on the spine of the hill and continued to appear on the beach without taking out the beautiful beech trees. Similarly, we were constrained in terms of future orientations to the land. We couldn't fight industries like fracking until we proved they had caused bodily harm. Could such ever be "proven"? Ecocritic and material feminist Stacy Alaimo considers materials and the contexts from which they come to be more fluid than we might suppose, informing the many ways we make sense of space and ourselves in a kind of complex social and environmental bodily flux across scales of place and time. Indeed, she acknowledges agency of the materials themselves.[14] Materials might move through human bodies, telling a tale that some would rather suppress. In this case, the material moved through a ravine, and the ravine moved through the discarded material.

I believe this acknowledgment of those complex interactions is the first step in determining what is sustainable. Seeing the processes in the Outdoor Classroom allowed me to acknowledge the complexity of the adjunct system. Adjuncts could never be disconnected from a larger system; to do so might topple the "ivory tower" of the contemporary university system. Yet, increasingly, they must be acknowledged and valued, for this system is built largely on the labor of adjuncts.

Restoration

Across from the stone sign that led to the Boyce Mayview Park, a road turned steeply below the park. I was to learn that it led to the Wingfield Pines Conservation Area, a place that would be the site for many lessons for the Outdoor Classroom. The area, a flat valley that opened to a series of ponds emptying into Chartiers Creek, had a different history than the Mayview State Hospital area. My impression of the area on my first visit in the spring after I'd been hired was that it was merely a paved-over abandoned swimming pool and a weedy parking lot.

The land is now part of the Allegheny Land Trust, but it was never farmland like the areas near the Outdoor Classroom above it. It is a floodplain; water collects from the surrounding hills and empties into these bottom wetlands. Once stigmatized because it could not be farmed, it was strip-mined in the mid-1900s, the Pittsburgh Seam coal being "just a few feet below the surface." Such strip mining involved stripping the topsoil to get at the coal and resulted in a "series of long, narrow ponds separated by low ridges of overburden." Areas like these were eventually labeled as "Abandoned Mine Drainage" (AMD) sites due to the high concentration of acidity, heavy metals, and sediment.[15]

Interestingly, industrialization gave way to recreation. John Oyler, a local news writer, recalls the ponds in the 1940s: "Blue Ponds, the site that is now called Wingfield Pines, was a wonderland of nature for those of us growing up in our neighborhood—a great place to hike and fish and enjoy the wildlife there." Perhaps this bent toward recreation explains later attempts to build a golf course and swimming pool and a newer swimming pool with a bathhouse from the 1960s through the 1990s. Ultimately, these endeavors proved unprofitable and ceased by the late 1990s, leaving behind the weedy parking lot. In 2001, when management of acid mine drainage became an acknowledged concern, the Allegheny Land Trust acquired the area and "reinvented" it as the Wingfield Pines Conservation Area. A passive remediation system was put into place to resolve the large amounts of ferrous iron that turn to rust when they come into contact with Chartiers Creek. The remediation plan included "a series of settling ponds and wetlands" interspersed "*before* they reach the creek" [italics original]. That attempt to capture iron sediment was realized through a series of ponds and wetland filters.[16] At the Outdoor Classroom, we could walk through the series with students, testing oxygen levels and acidity and counting species from the first pond, which was red with oxides and contained very little life, to the fifth pond, which was brimming with life. On my first visit, we walked along the length of the ponds, from the initial "dead" red pond to the "living" ecosystems where, by the end of our walk, a pond teemed with frogs, frog eggs, birds, snakes, and plant life.

At the same time the space was "remediated," a new sense of identity and even value of this space emerged: it acquired a sort of dignity. When we looked up at the top of the ridge of the valley, we could see shopping plazas on the ridge above us. Yet at the bottom of the valley, we were in a strange little Eden that offered flood protection from heavy rains, soaking up water that would have otherwise barreled into Chartiers Creek and Bridgeville.[17] The deposit into Chartiers Creek was not far from the beach where we explored micro- and macro-invertebrates. Here was an attempt to value the place for itself, before it moved into larger worlds. I could not help but make a connection between this metamorphosis and the college classroom. How might universities value instructors and students who will eventually move in larger contexts? And while "remediation" can be problematic in terms of coursework and the university, in the case of Wingfield Pines, it meant simulating what had existed before disruption, and doing so in a way that valued and revived natural systems.

I was deeply concerned, however, that even as the land was treated for one past industrial practice, new fracking sites were emerging as issues of environmental justice concern, once again calling into question who and what could identify a landscape. Was it a land for restoration, or a land of production? Who decided how to "value" this land? Dryzek critiques projects such as fracking, wrapped in the guise of weak sustainability through "clean" technology, as a poor example of ecological modernization, or an ideology in which "management" of resources trumps nature's own agency. Dryzek refers to Maarten Hajer's notion of "'techno-corporatist' ecological modernization, which treats issues in technical terms, and seeks a managerial structure for their implementation."[18] These concerns are allied with those of environmental justice or just sustainabilities.

This awareness of greenwashing heightened my concerns regarding the research agendas of research-intensive universities. I was keenly aware, for example, that fracking largely developed at my own alma mater, Penn State, as an "alternative" or "bridge" energy source. But the reality is that fracking is merely a new means of fossil fuel extraction. I know, too, that much of the hydraulic fracturing research agenda does not address critical reflections about social and environmental justice in an approach to the environment. The land surrounding the Outdoor Classroom bore witness to varied social arrangements that are impacted by, then continue to impact space, and recursively, new social arrangements and orientations towards what is seemingly "natural."

In juxtaposing the Outdoor Classroom with the university adjuncting system, which also struggles with issues of social justice, I considered a more radical notion of deeper sustainability and resilience. It is one that honors listening to students, to faculty, to land, that goes beyond management of "resources" to perpetuate a given system. Instead, I considered more complicated and integrated ways of offering people and place dignity. Lessons at the Outdoor Classroom were always augmented in surprising ways by living systems around us, by the interpretations of, and insights from, students. And students often saw and shared things the instructors had missed. In "Sustainable Service-Learning Projects," an exploration of composition and service learning in communities, Ellen Cushman suggests that the instructor and the student become "collaborators," thus discussing the importance of reciprocity and "collaborative inquiry."[19] The Outdoor Classroom, situated so firmly in "place," allowed us to avoid "telling" students what to say; students were already telling us what needed to be said in interactions between student and teacher.

I left the Outdoor Classroom to return to the university in the fall with mixed feelings. The "casual" position did not sustain me monetarily much more than an adjunct position did, and it was quite far away from my home, so I felt guilty about expending the fuel for the drive. Ultimately, I accepted funding as a full-time graduate student, which came with its own monetary limitations. However, the Outdoor Classroom experience emboldened me to recognize both the materialities and ideologies of the university that are often invisible. Too, the Outdoor Classroom allowed me to think more deeply about learning and praxis, leading me to explore service-learning opportunities and to consider how to sustain this kind of engagement in praxis in writing courses. Students in my recent professional writing course made recommendations for sustainability initiatives on campus, and I was moved to explore ethics and sustainability in business courses.

Such rethinking also led me to begin to work with other graduate students and adjuncts to act with agency, to provide professional support for each other in spaces the university typically does not provide for adjuncts or graduate students. I returned to the university to attempt to trace the complexities of the systems for deeper critical understanding of sustainable and resilient universities. This collaboration has resulted in a cross-disciplinary communication platform in which staff, faculty, and students might share campus sustainability initiatives. It is a small start, but it is a start.

In my short time at the Outdoor Classroom, "place" had become established more quickly for me than during several years at a university. I can tell you where the poison ivy grows. But I can also tell you where there is a stand of sassafras, and where the blue heron soars. I can tell you about the fish that I saw in the fifth pond at Wingfield Pines, a miracle due less to humans than to the integrated forces of nature that have their own messages of agency and dignity that humans would do well to contemplate. That knowledge is just as important to me as any found in a book.

NOTES

1. Daniel Kovalik, "Death of an Adjunct," *Pittsburgh Post-Gazette*, September 18, 2013.
2. David Orr, *Ecological Literacy: Education and the Transition to a Postmodern World* (Albany: State University of New York Press, 1991), 92.
3. Derek Owens, quoted in introduction to Peter Goggin, ed., *Rhetorics, Literacies, and Narratives of Sustainability* (New York: Routledge, 2009), 8; Orr, *Ecological Literacy*, 23; John S. Dryzek, *The Politics of the Earth: Environmental Discourses*, 3rd ed. (New York: Oxford University Press, 2013), 146; Goggin, *Rhetorics*, 3.
4. Rob Nixon, *Slow Violence and the Environmentalism of the Poor* (Cambridge, MA: Harvard University Press, 2011).
5. Marilyn M. Cooper, "The Ecology of Writing," *College English* 48, no. 4 (April 1986): 368; Rachel Tillman, "Toward a New Materialism: Matter as Dynamic," *Minding Nature* 8, no. 1 (January 2015): 32, http://www.humansandnature.org/filebin/pdf/minding_nature/january_2015/Towarda_New_Materialism.pdf.
6. Sherman Cahal, "Mayview State Hospital," *Abandoned: The Story of a Forgotten America*, 2016, http://abandonedonline.net/locations/hospitals/mayview-state-hospital/.
7. Ibid.
8. Deborah Brandt, "Sponsors of Literacy," *College Composition and Communication* 49, no. 2 (May 1998): 169.
9. "The Outdoor Classroom," The Outdoor Classroom, http://www.theoutdoorclassroompa.org/.
10. Katrina Brown, "Lost in Translation? Resilience Ideas in Science, Policy and Practice," presentation, University of East Anglia, Norwich, England, September 17, 2009; Dryzek, *Politics of the Earth*, 159.
11. Ibid.
12. David Barton, *Literacy: An Introduction to the Ecology of Written Language*, 2nd ed. (Malden, MA: Blackwell Publishing, 2007), 32, 50.

13. James Porter, "Intertextuality and the Discourse Community," *Rhetoric Review* 5, no. 1 (Autumn 1986): 35.
14. Stacy Alaimo, "Trans-Corporeal Feminisms and the Ethical Space of Nature," in *Material Feminisms*, ed. Stacy Alaimo and Susan Hekman (Durham, NC: Duke University Press, 2007), 238.
15. John Oyler, "The Wingfield Pines Has Been Used for a Variety of Things," *Tribune Live*, November 13, 2013, http://triblive.com/neighborhoods/yourcarlynton/yourcarlyntonmore/5018173-74/wingfield-bridgeville-pines#axzz3jqxNk3Kt; Pennsylvania Water Science Center, "Coal-Mine-Drainage Projects in Pennsylvania," U.S. Geological Survey, last modified October 26, 2010, http://pa.water.usgs.gov/projects/energy/amd/.
16. Oyler, "Wingfield Pines."
17. Ibid.
18. Dryzek, *Politics of the Earth*, 173.
19. Ellen Cushman, "Sustainable Service Learning Programs," *College Composition and Communication* 54, no. 1 (September 2002): 42.

BIBLIOGRAPHY

Alaimo, Stacy. "Trans-Corporeal Feminisms and the Ethical Space of Nature." In *Material Feminisms*, edited by Stacy Alaimo and Susan Hekman, 237–264. Durham, NC: Duke University Press, 2007.

Allegheny Land Trust. "Wingfield Pines Conservation Area." Last modified April 14, 2016. Http://alleghenylandtrust.org/green-space/wingfield-pines/.

Barton, David. *Literacy: An Introduction to the Ecology of Written Language.* Malden, MA: Blackwell Publishing, 2007.

Brandt, Deborah. "Sponsors of Literacy." *College Composition and Communication* 49, no. 2 (May 1998): 165–185.

Brown, Katrina. "Lost in Translation? Resilience Ideas in Science, Policy and Practice." Presentation. University of East Anglia, Norwich, England, September 17, 2009.

Cahal, Sherman. "Mayview State Hospital." *Abandoned: The Story of a Forgotten America.* 2016. Http://abandonedonline.net/locations/hospitals/mayview-state-hospital/.

Cooper, Marilyn M. "The Ecology of Writing." *College English* 48, no. 4 (April 1986): 364–375.

Cushman, Ellen. "Sustainable Service Learning Programs." *College Composition and Communication* 54, no. 1 (September 2002): 40–65.

Dryzek, John S. *The Politics of the Earth: Environmental Discourses.* 3rd ed. New York: Oxford

University Press, 2013.

Goggin, Peter. Introduction to *Rhetorics, Literacies, and Narratives of Sustainability*, ed. Peter Goggin, 1–12. New York: Routledge, 2009.

Nixon, Rob. *Slow Violence and the Environmentalism of the Poor.* Cambridge, MA: Harvard University Press, 2011.

Orr, David. *Ecological Literacy: Education and the Transition to a Postmodern World.* Albany: State University of New York Press, 1991.

Pennsylvania Water Science Center. "Coal-Mine-Drainage Projects in Pennsylvania." U.S. Geological Survey. October 26, 2010. Http://pa.water.usgs.gov/projects/energy/amd/.

Porter, James E. "Intertextuality and the Discourse Community." *Rhetoric Review* 5, no. 1 (Autumn 1986): 34–47.

Tillman, Rachel. "Toward a New Materialism: Matter as Dynamic." *Minding Nature* 8, no. 1 (January 2015): 30–35. Http://www.humansandnature.org/filebin/pdf/minding_nature/january_2015/Towarda_New_Materialism.pdf.

Meeting across Ontologies: Grappling with an Ethics of Care in Our Human-More-than-Human Collaborative Work

Bawaka Country (including Laklak Burarrwanga, Ritjilili Ganambarr, Merrkiyawuy Ganambarr-Stubbs, Banbapuy Ganambarr, Djawundil Maymuru, Kate Lloyd, Sarah Wright, Sandie Suchet-Pearson, and Paul Hodge)

> *Wetj* reminds us that everything is interrelated, that things can't be separated out for convenience or comfort.
>
> —Bawaka Country

> LAKLAK: It is important to share because it is the Law. It is the system from the old people, going back thousands and thousands of years, forever. This system has been standing really strong. So our mob, we grow up, and we see those rules. We have been told, "You must not be greedy. You have to share with everyone."
>
> We have been told, and we have the knowledge in our heads, what we have to share with the family. It is a cycle of sharing. You share with some; others share with you. It doesn't matter if you're in the city or wherever; we always share. That's *wetj*: sharing and responsibility.[1]

Our research collective first explicitly talked about *wetj* on the beach at Bawaka, an Indigenous Yolŋu homeland in Northeast Arnhem Land, Australia, in 2010. We were collaboratively writing a book about Yolŋu mathematics—a book that was aimed at discussing the underlying patterns and connections that underpin Yolŋu ways of being and connections with each other and a more-than-human, sapient, and sentient world. At the time, we were working on a chapter on counting and division. Laklak, a Yolŋu elder and senior sister of the collaboration, pointed at the chapter title written on a page of the notebook: "This is about *wetj*," she said.

Wetj is a foundational Yolŋu concept. It talks of the importance of sharing, the responsibility to share. It places all, human and more-than-human, within a web of kinship and connection, and it links us together through obligations of attention, response, and responsibility. So that, as well as counting turtle eggs on the beach that day or any day, as well as dividing them up, we were realizing *wetj*, actualizing our kinship with our extended human and more-than-human family—the turtle, the beach, our ancestors, the songs and stories, and all the myriad human and more-than-human relationships that co-create Country.

An Aboriginal English term, Country refers to a specific place or homeland and encompasses the human, more-than-human, tangible, and intangible forces that shape, create, mutually care, and become together in, with, and as place/space. Thus Bawaka Country refers to the diverse land, water, human and more-than-human animals (including the human authors of this chapter), plants, rocks, thoughts, and songs that make up the specific Indigenous homeland of Bawaka.[2]

Wetj is everywhere. It is everything. *Wetj* holds people and Country. It is ethics; it is becoming together; it is responsibility. As a research collective seeking to work within a Yolŋu ontology, *wetj* underpins our work, informs our decisions, brings us into being together. Yet *wetj* does not sit easily within the research norms and frameworks promulgated by neoliberal universities that value auditing processes and individualized, competitive accountability above all else.[3] *Wetj* demands different modes of responsibility, different ways of being and becoming together; it demands that we work, be, and co-become differently.[4] It is this dilemma, these ontological challenges around knowledge and authority as well as practical considerations around accountability and reciprocity that we focus on in this chapter. In doing so, we use *wetj* and an ethics of care to make visible ontological disparities as well as to show the openings and opportunities that can come from highlighting such failures, rejections and mistakes.[5]

We write this chapter as an Indigenous and non-Indigenous, human and more-than-human research collective that has worked together since 2006. In doing so we talk about our work together with and through *wetj*. We discuss some of the challenges we face as we come together from diverse worlds—as an elder and as caretakers of Bawaka Country, as an Indigenous principal and as a senior teacher at a community school, as non-Indigenous academics at two universities, as Yolŋu running an Indigenous led and owned cultural tourism business, as members of families, and as part of Bawaka Country. As a coauthored piece, the main text represents "our" collective Indigenous and non-Indigenous, human and more-than-human voice, developed through discussions together, writings passed between us, and our shared experiences shaped and enabled by Bawaka Country. However, within the chapter at various points, different human authors share their specific experiences and perspectives. These mini-stories are indented and the human authors specified.

Wetj and an Ethics of Care

Wetj means to share, with humans and more-than-humans, as kin, through specific relationships that tie us all to each other and to Country. It is through these relationships, these responsibilities, that we exist, that we *become*, so that *wetj*, ultimately, is about becoming together, *co-becoming* through kinship.

> LAKLAK: For us, kinship lies at the heart of everything. A world with no kinship is a world that does not have a true existence. Kinship gives everything meaning, order, balance. *Wetj* tells us about kin, about respect and duty, about who is close and who is distant. Sometimes we share to make those who are far away come close. It is important to our identity as Yolŋu.

To *become together* as a relational and more-than-human practice of sharing is to begin to understand *wetj* as an ontology of co-becoming. *Wetj* encompasses the myriad processes, actions, feelings, and connections that constitute the entangled co-becoming of "all humans and non-humans, actors, actants, everything material, affective."[6] *Wetj* allows and enables as it underpins our collective existence through relationships. To attend to the digging of *ganguri* (yams) or collecting, counting, and sharing of *miyapunu mapu* (turtle eggs) at Bawaka is to attend to *wetj*; as we gather, share, talk, and eat, we are partaking in a Yolŋu ontology of co-becoming together. And it is this relational connectedness—embodied in the concept of *wetj*—that is central to our Indigenous and non-Indigenous, human-more-than-human collaboration and our efforts to care for each other as Bawaka Country cares for us.

This co-becoming, as we share and care together, resonates with Victoria Lawson's care ethics and social ontology of connection; Jeff Popke's ethics that comes into being *through* knowing, doing, and being; and Deborah Bird Rose's embodied responsiveness via an ethics of attention.[7] Each of these formulations of care reveals an explicit attentiveness to the actual practices of relationality with all their propensities and affects. For us, it is this attentiveness in the human-more-than-human realm that underscores our ethics of care as we embrace *wetj* and become together. It is the very essence of the relationships attained through Yolŋu sharing.

Authority and Knowledge

A Wondrous Mind

Everything is touched by *wetj*. *Wetj* is practical, informing day-to-day issues and decisions of the most seemingly prosaic kind. Yet *wetj* is ontological. It speaks to deep Law, to the very meaning of what it means to be human and to live in this world. Indeed, *wetj* denies many separations that underpin work in a mainstream university setting. It denies a distinction between the prosaic and the profound, between work and family life, between weekdays and weekends, between humans and more-than-humans, between now and later.

The neoliberalization of the university, in Australia and internationally, is but one manifestation of a uniquely Western European historical trajectory. Universities are seen to uphold and reinforce traditional authoritarian assumptions regarding the location and deployment of rational knowledge, and to endorse the academic as the distant "expert" or "knowledge broker" where dissemination of knowledge flows from the academic to the community. Knowledge and truth have long been, and still are—despite long-running, profound feminist, cultural, and other critiques—seen as absolute: things must be right or wrong, rational or irrational, while scientific knowledge is understood as disembodied, universal, and placeless. Such approaches deny, dismiss, and make invisible a myriad of vastly diverse ways of knowing and being.[8]

As such, despite their being set up as places of learning and wonder, we find that the ontological arrogance of many Western university institutions in fact blocks, dismisses, or belittles the "wondrous mind" that Merrki here describes so poignantly:

MERRKI: I reckon not only politically but through everyday life there are people who don't look at things our way. There are people who need to understand more, to be educated, to have a broad mind, to accept things as they are and to work their way around if they don't like it. That is what I think—a healthy person has a mind full of wonder. A wondrous mind learns so much more than a mind that is sick. A narrow-minded person is a sick person.

Working in an intercultural and more-than-human research collective encourages us to find our wondrous minds. And there is deep *wetj* in the way Laklak, Ritjilili, Merrkiyawuy, Banbapuy, Djawundil, and Bawaka Country have generously shared their insights and knowledge with the collective's non-Indigenous members. Nurturing our wondrous minds leads us to the edge of what *is*, as well as what is known—that is, to ontology. Indeed, coming and becoming together and becoming family lead to moments of connection and deep insight, as well as moments of uncertainty and puzzlement. For as we seek to nurture our wondrous minds, we all face moments of "sickness" both within and without. Ongoing recognition of the ways that deep ontological differences, different experiences of colonization, and the realities of racism flow through our research collective can be challenging, frustrating, *and* transformative as we work where we can within antiracist and anticolonizing practice.

Not Lone (Human) Wolf

The non-Indigenous researchers came into our collaboration as newborn babies in terms of their knowledge of Yolŋu ways of being and knowing the world, and, despite learning so much, they are still very much in their nappies.[9] However, some of the learning that has occurred has gone to the heart of academic practice—practice that historically has nurtured an individualized (hu)man self, a "lone wolf" academic. Part of forging anticolonizing practices that challenge academic conventions has been to make explicit the concept and practice of family as central to what it is to become together. Highlighting the ways "loved ones actively and constantly shape the type of research we want to do" and ultimately play a formative role in the knowledge produced through their "physical presence" shifts the individual-centered nature of academic research.[10]

As the following two stories highlight, the academics have started to grapple with what was unimaginable to them ten years ago, and in doing so they have had to overcome not only practical obstacles and challenges but also—and perhaps most importantly—blockages in their own ways of thinking.

> MERRKI: How is it that we make this happen? We should talk about how we all have families, we all have children, because one of the things that is important for us is that we are all women. This is what helps us meet challenges. Our understanding of each other is important too. Sometimes we don't even talk to each other but we understand each other. We do things and it works. That is something that is different that is going on. Sometimes we don't even tell each other what we are doing; we just do it without planning. Like with the kids. One thing is that we always do everything for the kids first before we do it for ourselves. We make sure that they are understanding, their needs are met, through all this. It can be a mayhem sometimes, but in the end everyone is feeling good. Everyone is satisfied. That is one of the things that can be in working together. And it is so wonderful that our husbands understand us, the balance. I think we have met the husbands that every woman should have. We are lucky to have these men in our lives.

Kinship relationships are fundamental to *wetj*, and therefore attending to our human family has been a point of connection for all of us. However, being able to embrace the social networks that enable our work has been challenging in the context of the university. We have not yet acknowledged our husbands or children as authors of our work—despite the ways they've shaped it so intensely. The place of Indigenous families and more-than-human kin has been integrated to some extent into authorship and grants (see below), as well as in applications submitted to the universities' human research ethics committees, but non-Indigenous families continue to be largely silenced in those same forums. We have not applied for or used grant money to pay for airfares for non-Indigenous family members, or to pay them for their time and effort, while children, both Indigenous and non-Indigenous, are only ever partially recognized. In this case, it is non-Indigenous husbands and children whose roles are behind the scenes, making the fire, caring for children either with us at Bawaka or left at home down south as airfares are too expensive for them to attend. Western constructions of nepotism and assumed boundaries between work and home have been as yet too strong to overcome.[11]

However, we have been able to shift the boundaries of authorship in other ways, in particular responding to the *wetj* imperative to share with our more-than-human kin.

Sandie: We academics always felt an imperative to acknowledge the level of authority contributed by our Indigenous collaborators through coauthorship of articles (through guidance, access, and of course sharing and coproduction of knowledge), although even this has shifted as more of our Indigenous colleagues join us as coauthors. However, what we had never considered prior to our work together in Bawaka was that we were in fact working with and as Bawaka. That in taking seriously the active agency of Country, we were not only a human collaboration but a human-more-than-human group. As such we realized, as an imperative of *wetj*, that we need to acknowledge Bawaka Country as an author of our work. We initially started authoring as the human authors with Bawaka Country as an additional author. We quickly realized we weren't in fact separate from Country, and, indeed, it was Bawaka Country, including us human authors, that was shaping and enabling our work. What was an immense ontological shift for the academics was met with bemused astonishment by the Bawaka mob who felt that recognizing and celebrating Bawaka's contribution was an extremely obvious move to take.

As Merrki adds, "She—Bawaka—is the author; we—humans—are only the voices. And we are part of her language."

When we put forward this authorship to our journal publishers (initially for articles that had already been accepted), it was with some trepidation. However, to our surprise, it has been accepted and indeed embraced enthusiastically. Micro-issues have emerged—for example, linking the human-more-than-human authors with that word *including* proves particularly difficult for publishers, web interfaces, and typesettings (and it hasn't always been successful); similarly, registering for a conference and needing to state one's position and title, even email address, for Bawaka Country can be a challenge. But we have been impressed with the ways people have helped us to overcome these challenges once the initial ontological/imaginative hurdle was shifted.

Playing the Game

Recognizing and shifting such ontological and practical barriers is incredibly exciting work. However, there has also been the need to acknowledge and work with the assumptions and needs of the university system to ensure our research and collaboration continues to be supported, recognized, and funded. Metric-based accounting regimes (in Australia, the so-called "Excellence in Research Australia," or ERA) now dominate the neoliberal university, determining the allocation of resources through annual reviews and representing what Annelie Bränström-Öhman calls a shift from "content to counting." Like Bränström-Öhman, the Great Lakes Feminist Geography Collective (Mountz et al.) and Rachel Pain consider how this shift actively shapes the way academics work and the outputs they produce; likewise, Noel Castree argues that the research assessment systems that form part of this counting define academic "value" in rather narrow ways. Don Brenneis, Cris Shore, and Susan Wright highlight the significant role universities play in cultural reproduction, and argue that the shift towards an audit culture means that disciplines such as anthropology are increasingly open to reevaluations of what "counts" as valid or "robust" knowledge. Their concern is that as the politics of accounting takes hold, the knowledge that is seen to be legitimate within academia will become increasingly narrow and limiting.[12]

Community-based researchers must answer to institutional metrics and norms alongside the demands of the research they are undertaking and the expectations and agenda of their community partners and family members, as dictated through cultural obligations. As Sue Jackson and Louise Crabtree reflect, "Australia's research evaluation framework does not recognize the value of this plurality [the multiple motivations for, and benefits or outputs of, research], nor does it ensure it is reflected in its measures of research quality and impact and in the systems of support helping researchers to capably engage with the community sector."[13] There are also different accountabilities at play, under *wetj*, that work with different norms, values, assumptions, and worldviews. Recognition of authorship, for example, should be based on seniority within a Yolŋu world, while decisions as to who may share, translate, and (re)tell stories are based on *Rom*, on Yolŋu Law.

Both Indigenous and non-Indigenous members of the team, then, balance different accountabilities and imperatives. The academics constantly work under the pressure of producing academic outputs that meet criteria of research excellence, such as books, peer-reviewed journal articles, book chapters, and competitive grant or consultancy income. Yet they do so within the context of a broader collaborative aspiration of reciprocity, a response to *wetj*'s imperative to share, and a desire by all involved in the collaboration to produce mutual benefits, not only different things that achieve multiple aspirations but also single outputs that can be of benefit in different ways. Yolŋu members also work as teachers and tourism providers and have cultural obligations to share knowledge appropriately.

Respecting this obligation to share and drawing inspiration from postcolonial and feminist work, our collective always tries to communicate openly to find creative ways to make our research work in our respective contexts.[14] One strategy is to discuss our needs and desired outputs as openly as possible, though some cultural understandings go beyond easy description. As such, when the idea for our second book was born, we agreed immediately that its audience was a wider popular readership as desired by the Yolŋu women—but we all also wanted it to be seen as an academic publication to reach an academic audience and contribute to the academics' research outputs as desired by their universities. With these desires in mind, we were delighted to have a top publisher take the book on. In composing the book, we included specific things to appeal directly to an academic audience—for example, a list of references for further reading—even as we used an accessible style that could be appropriate for visitors on Yolŋu-led tours as well as academics and students.

When the book was published, Macquarie University categorized it initially as an "A1 publication"—an authored book by a commercial publisher, the best classification for academic publications in determining kudos and government funding. However, the academics then were informed that, due to faculty intervention, it had been reclassified "J1"—a problematically less prestigious category encompassing major creative works. After further investigation, they were told that because faculty could not see academic citations or footnotes throughout the book, they felt it was not of scholarly import. Thanks to an incredible support network of academic mentors, the academics put together a case explaining the scholarly importance of the book and its legitimate reliance on primary research, a case helped by the University of Newcastle's prior acceptance of the book as A1, and it was then recategorized as A1. This example illustrates, however, a lack of understanding within university structures, and their struggle to recognize appropriately other ways of knowing and to reward engagement and the production of "less traditional" outputs with community partners.

Due to institutional assumptions about the generation of knowledge, our collaboration has hence bumped up against boundaries of what knowledge can be counted, and by and for whom. Yet, "What if we counted differently?" Alison Mountz et al. ask. "Instead of articles published or grants applied for, what if we accounted for thank you notes received, friendships formed, collaborations forged?"[15]

Rejection and Failure

Importantly, *wetj* means working for the long term, an antithesis to short-term funding cycles and the difficulty of gaining tenured employment in the contemporary university. *Wetj* is, as Laklak explains in the introduction earlier, a "system from the old people, going back thousands and thousands of years, forever." It means building relationships that are more than long-term; they are intergenerational. *Wetj* is a way of being in the world. It means that relationships are relations of kinship, not of short-term convenience and not to be entered into or taken lightly. Yet to make such long-term commitments is something of an anathema within the contemporary university setting. Kate reflects on the process that brought the academics to Bawaka.

> KATE: Before we made our connection with Bawaka Country and the Burarrwanga family we had some false starts. Some Indigenous groups we approached elsewhere in the Northern Territory weren't interested in working with us while others were, but then situations changed for them. At the time these rejections and failed relationships were the source of much angst. But for any collaboration to be real, people have to be able to say no. We see them as important steps in our journey to develop strong, long-term research collaborations with Bawaka Country and the Burarrwanga family. And it is a long-term thing. But we know that a neoliberal university doesn't tend to work on long-term timelines, and many people—those on fixed-term contracts, or PhD students pushed to complete in three years—have an uphill battle to find ways to make long-term relationships work.

Kate, Sandie, and Sarah, as tenured academics, recognize that many barriers exist, which means that other researchers cannot enjoy the opportunities they have had in establishing and nurturing our collaboration. Indeed, their work has been possible only due to their privileged positions within the academy—continuing, tenured positions that include hard-won maternity-leave conditions (further discussed later in this chapter); the security to make longer-term commitments; and the ability to utilize internal university grant opportunities. Scholars mrs kinpaisby and Deirdre Conlon, Nicholas Gill, Imogen Tyler, and Ceri Oeppen refer to these longer-term research relationships as "slow research" and maintain that long-term commitments are essential to the development of good participatory practices. Yet the neoliberalization of the university and its concomitant casualization of the work force have put academics in exploited positions in which they cannot make or support the sorts of long-term research commitments at the heart of our collaboration. Indeed, Aline Courtois and Theresa O'Keefe see casualization as "both a consequence and an instrument of neoliberalization, making resistance difficult and paving the way for a complete reorganization of the sector along managerial, neoliberal lines."[16]

Accountability and Reciprocity

Wetj requires that we always try to ensure that our research collaboration is founded on common priorities of trust, reciprocity, relationships, and shared goals. *Wetj* sees all beings and things as inherently connected and requires different ethical understandings of processes and outcomes. This is not always easy, and we have faced a range of interpersonal/intercultural and institutional challenges in the areas of accountability and reciprocity.[17]

Money, Money, Money

Sandie, Kate, and Sarah: Administration of internal and external grant money has been tricky due to the rigid nature of university administration systems and their inability to take into account alternative systems of accountability. Paying senior knowledge holders and community informants for their time is not as simple as processing an invoice. Issues of communication, documentation, paperwork, and financial systems are all involved in a complex and messy interplay. Often we need to pay people up front in cash and then claim from the university—a seemingly unproblematic process relying on cash advances, but a process that in reality requires careful planning and navigating of complex university financial systems.

When we were awarded a prestigious Australian Research Council grant in 2013, we were so excited because the budget included substantial funds for paying Indigenous co-researchers. But managing the transfer of money from the grant has been and continues to be fraught, due to different frameworks and ontologies. Laklak has always guided the discussion with her sisters on who should get paid what and how. The family decided that vouchers for the local supermarket was an effective and efficient way to pay people for a large portion of their time (or at least the best option of a tricky lot). However, knowing when to send vouchers,

> to whom to send the vouchers, and responding to contingencies that inevitably arise means that the academics, who "control" the finances, feel like they are often in the role of colonial administrator. Although it was agreed by all that this is the most effective way to "manage the money," and it works much better in comparison to other methods we have tried (such as waiting two months for an invoice to be processed when money is needed immediately to pay for food), the academics still feel uneasy about the voucher system. Luckily, despite the frustrations with the financial systems, we have developed connections with administrators (who support both the academics in the university and the Yolŋu in Arnhem Land) who understand the different circumstances and work as closely with us as possible to enable smooth practices and outcomes.
>
> Negotiating these "financial transactions" is not only an administrative, bureaucratic hurdle, but in many ways a personal challenge. It has pushed us to acknowledge consciously our identity and relationship with money as coming from a Western middle-class ethic. We are very conscious that we are paid regularly, securely, and well for our work—and that we are succeeding in that work in large part due to our research collaboration with Bawaka. Working with Laklak and her family, many of whom do not have a secure or high income, we have had to rethink and challenge our perceptions and personal anxieties around money as well as separations between work and home. For example, our "personal" funds are often drawn into the *wetj* system of sharing "outside" of our "research" transactions. While we have worked hard to shift our (and our families') cultural frameworks as we relate to our Indigenous collaborators as interrelated family, it has not been so easy for the university administration systems.

There are different layers of accountability and different sets of ethics at work, based on fundamentally different ontologies, and it sometimes can be hard to align these categories.[18] For funding bodies, for example, the ultimate accountability is to do what you say you will do in a grant application and spend money aligned with the budget. But within Yolŋu worldviews, actions must be underpinned by *wetj*, the need for reciprocity and responsibility with human and nonhuman kin and *Rom*. So that means matching up ethics—or, put another way, translating things within different frameworks. For the Australian Research Council, we wrote in payments to traditional knowledge authorities at a rate set by the Australian Institute of Aboriginal and Torres Strait Islander Studies. For the family, we pay senior people as part of *wetj* and out of respect. In some ways, it amounts to the same thing, but the ethical framework underpinning it is different. Generally, it is the academics who write the wording of grants and enumerate the details of budgets, after discussion of concepts, plans, and budgets with all of us. This process takes time, and it means tight deadlines might not be workable. We all have to be very careful about what goes into grants to make sure we are not going to find ourselves in an impossible situation, yet we always need to factor in flexibility. It means simply not going for some things or writing applications in ways that might not give us the best chance of getting the money. But it is a lot better than having money and not being able to work in ways supported by *wetj* and *Rom*. Our next challenge will be to write Bawaka Country into the budget for our grant applications. If we pay people, we should have money set aside to support Bawaka Country too. We'll see how that goes!

A Time and Place

Wetj reminds us that everything is interrelated, that things can't be separated out for convenience or comfort. For example, politics underlies everything we know and do, although sometimes in subtle ways. And sometimes we're concerned that it is *too* subtle. We feel angry about things that happen, like former prime minister Tony Abbott's comment about Indigenous connection to Country being a "lifestyle choice," a comment that we argue is based on deeply offensive assumptions about people and place. But as a collective we try to talk back in positive ways—talking back about the deep connections between people and Country, the way that people are in fact part of Country, and the impossibility that such co-constitution could be just a "lifestyle choice." These deep connections are about the very meaning of the world and our place in it. To talk the way Abbott does reveals long-held conservative views that social circumstances are the responsibility of individuals, as if the government is somehow no longer responsible at all, and lays the groundwork for ongoing rounds of dispossession and ontological violence.[19] But whether and how our response is picked up, well, sometimes we feel it is too easily ignored.

> SARAH: We wonder if some of this ignorance comes from romantic ideals of Indigenous peoples and an inability for people to embrace difficult and confrontational interpersonal/cultural situations. For example, tourists who have gone to Bawaka have often expressed feeling uncomfortable when politics or stories of oppression and dispossession are discussed. One tourist talked about the discomfort that came out of a discussion on the Northern Territory Intervention. She said that when the

focus was on connection to Country there was a sense of "incredible ease," but that when it became political they all fell to pieces; they didn't know what to do in those spaces. She somehow couldn't see that connection to Country *is* political—it is unavoidable. However, there is also a political imperative to put a positive spin on issues, to focus on strengths in order to counter the media bias and negative stereotyping of Aboriginal people. Yet we worry that people read a romanticism into our work and learning together, and we are also conscious that it can lead to some glossing over of certain issues—early mortality rates, sickness, racism, genocide, ongoing bullshit, sorry business, ongoing deaths. We have navigated this space collaboratively, although we all feel there is more we could do and worry about the problematic, "deep-colonizing" consequences of some of our work.[20]

We Owe a Lot

Wetj is always present; it holds both the past and the future in the present and links us (in the broadest more-than-human sense) to each other in infinite, connected ways through time, space, and place. Working through *wetj* means recognizing that our work together owes so much to people and Country who have come before and to people and Country who are still to come. Yolŋu elders of the past have passed on knowledge and kept Yolŋu knowledge alive through incredibly difficult circumstances. The agency of Bawaka Country and the Yolŋu ontologies that support it have resisted and survived. As our dynamic collaborative learning demonstrates, *wetj* has resisted colonial and current "post"-colonial assaults on autonomy and connection to land.[21] This ongoing learning—about responsibility, justice, place, and belonging—also encompasses and builds upon a range of movements for change both within and outside of academia, which have constantly challenged assumptions and improved practices.

Sarah: There is a lot of support for our work together, sometimes in surprising places. We know that much of this support is a result of ongoing struggles, of Indigenous peoples' movements, the feminist movement, and the union movement. We feel we owe a lot to the amazing people and the more-than-humans who work with those movements who have shifted the terrain. Indigenous rights struggles have opened up spaces for speaking back in academia and changed the way research is done in radical ways. And the National Tertiary Education Union of Australia has prioritized flexible work conditions and good parental leave for academics so

> that we have been able to continue our work in ways that would not have been possible in the past. So a huge shout-out to those movements—through multiple generations. Of course, everything is still so deeply colonized and so much more needs to be done.

Despite these positive gains, however, the power of the university, which supports our work in so many ways and yet remains unidirectionally accountable to a broader tax-paying "public" (as realized through government bureaucracies), remains a dominant force in our relationships.

Death by a Thousand Cuts

> SANDIE AND KATE: A year and a half after taking a first brief to the vice chancellor of Macquarie University to see if he would support us putting in a nomination for an honorary doctorate for Laklak, we finally got news that the University Council had resolved to confer upon her the degree. As brilliant and appropriate as it is, it was a very frustrating eighteen-month process, due not to any underlying conspiracy theories or explicitly wicked policies but mini/microscale bureaucratic, managerial blockages. From not having a clear nomination deadline or call for nominations communicated, we missed the first deadline to submit the nomination; having waited six months, we then found out the nomination did not proceed to the next stage, possibly due to the lack of a champion at the faculty level (something the nominators were not aware was needed); even after it made it to the final decision-making meeting, we would not have been aware of the success of the nomination if it were not for colleagues directly involved; and without those networks and connections, communication of the nomination might not have made it to Laklak, who was at the time in hospital. Terrible communication and incredibly poor management processes plagued what should have been a straightforward and celebratory process.

The small blockages that pop up every day are not really small—they add up to more. When we started talking about challenges to include in this chapter, we found it hard to actually pinpoint the big structural problems that face us. Instead, we kept on thinking of the enablers—how we are able to fight for articles to be published despite one reviewer who doesn't get it, or how we can find recognition here or there thanks to some amazingly supportive colleagues and administrative staff scattered around the universities and elsewhere, or how we help each other and love our collective so much. But the thing is, we understand that those big structural problems, the racism, and the deep colonizing processes, at the end of the day are made up by these thousands of "small annoyances" that just build and build and build. That's the death by a thousand cuts.

Working and becoming together as a collective with an ethics of care based on *wetj* really helps. We can take turns feeling disempowered, and turns being the one to lift us up to keep going; we can take care of each other and always learn from the care that Bawaka bestows on us.

> MERRKI: Every day anything can let you down, but if you have a mind that wants to learn, you can become. You still learn; you overcome those obstacles. You do things the right way, you look after your people, your Country, the way you want things to be. If you look after your mind, you solve problems.
>
> NANU (Laklak, Ritjilili, Merrki, and Banbapuy's granddaughter): This is how we go. We keep going, keep going. We tell stories.
>
> MERRKI: And you can, you can solve problems and work things out to make your life the way you want it to be.

Conclusion

In an era where neoliberal universities increasingly seek to count and measure, and dominant culture is led by an ontological arrogance that refuses to admit that other ways of knowing and being are possible, let alone legitimate or valuable, *wetj* demands something different. In reframing a discussion on the beach at Bawaka from one of counting, measuring, and dividing to one of *wetj*, Laklak shifted the ground. Yes, *wetj* is counting and measuring and dividing, but it is more. *Wetj* means that counting, measuring, and dividing are expressions of care, ethics, and obligation. As a collective, we have asked, is it possible to shift notions of research measurement and accountability in this way? How can our research become a matter of *wetj*? Of sharing and responsibility? How can we decenter "the human" in these relationships to recognize and respect co-becomings with more-than-human kin? Is there a possibility that *wetj* might underpin *all* research, that human-more-than-human sharing and responsibility might become the measure by which *all* research is understood and judged?

As a human and more-than-human collective, we often fail to measure up to the depth of *wetj*. Kate, Sandie, and Sarah, as non-Indigenous academics just emerging from their nappies, struggle even to understand the complexity of the term, to feel and live and understand themselves by and through it. Laklak, Ritjilili, Merrki, Banbapuy, and Djawundil also find themselves in intercultural worlds, forced to engage with school systems and a dominant non-Indigenous culture every day. It is in these difficult spaces—these colonized, neoliberal, and racialized worlds—that our collective tries to navigate a different kind of research ethics, a research ethics guided by Bawaka Country, underpinned by *wetj*, by an ethics of caring, sharing, and co-becoming. It is unsurprising that this has at times been a challenging task.

In this chapter, we have highlighted some of our many failures, rejections, and mistakes, and we have discussed some of the lessons we have learned as well as shared some of our partial successes. As Judith Halberstam explains, "Under certain circumstances failing, losing, forgetting, unmaking, undoing, unbecoming, not knowing may in fact offer more creative, more cooperative, more surprising ways of being in the world."[22] Through our failures we indeed have become a stronger collaborative team. And in meeting these challenges together, we have found joy and comradeship; we have become kin. This is the heart of it. We are embracing a system that insists that *wetj* is part of everything. We come to *wetj* from different places, and we navigate it in different ways. For *wetj* speaks to both possibility and to limitations, to closeness and to distance.

These reflections have been important to us, to help us see where we have come from and where we are going. They have reminded us of how we have been challenged by institutional structures, university culture, and metric-based accounting regimes, and how in turn we have challenged them. We have been reminded of the demands and tensions that both worlds place on us all as we work together. Through *wetj* we have been able to see some ways that an ethics of care can underpin our work, inform our decisions, and bring humans-more-than-humans into a more nurturing and sustainable becoming-together.

Our collaboration continues to challenge us to clear the blockages in our own ways of thinking. Ultimately, working within *wetj* is beautiful. It promotes an ethics of care that recognizes care, not between several discrete and divergent units, but between a human and more-than-human collective that emerges together, with potentially profound implications for intergenerational self-determination, sustainability, and well-being. We care as kin, as part of each other, and as part of Bawaka Country. We care, within *wetj* and as an expression of *wetj*, and so co-become differently in ways that nourish wondrous minds as we meet across ontologies.

NOTES

1. Laklak Burarrwanga and Family, *Welcome to My Country* (Melbourne: Allen & Unwin, 2013), 37.
2. For a detailed discussion of Country, see Bawaka Country (including Sarah Wright, Sandie Suchet-Pearson, Kate Lloyd, Laklak Burarrwanga, Ritjilili Ganambarr, Merrkiyawuy Ganambarr-Stubbs, Banbapuy Ganambarr, Djawundil Maymuru, and Jill Sweeney), "Co-Becoming Bawaka: Towards a Relational Understanding of Place/Space," *Progress in Human Geography* 40 (2016): 455–475, http://dx.doi.org/10.1177/0309132515589437; and Deborah Bird Rose, *Nourishing Terrains: Australian Aboriginal Views of Landscape and Wilderness* (Canberra: Australia Heritage Commission, 1996), https://www.environment.gov.au/system/files/resources/62db1069-b7ec-4d63-b9a9-991f4b931a60/files/nourishing-terrains.pdf. For discussion of the relationship of these conceptions of connectivity to sustainability in environmental education, see Laklak Burarrwanga, Meerkiyawuy Ganambarr-Stubbs, Banbapuy Ganambarr, Sandie Suchet-Pearson, Kate Lloyd, and Sarah Wright, "Learning from Indigenous Conceptions of a Connected World," in *Enough for All Forever: A Handbook for Learning about Sustainability*, ed. Joy Murray, Glenn Cawthorne, Christopher Dey, and Chris Andrew (Champaign, IL: Common Ground, 2012), 3–13.
3. See Aline Courtois and Theresa O'Keefe, "Precarity in the Ivory Cage: Neoliberalism and Casualisation of Work in the Irish Higher Education Sector," *Journal for Critical Educational Policy Studies* 13, no. 1 (June 2015): 43–66; Cris Shore and Miri Davidson, "Beyond Collusion and Resistance: Academic-Management Relations within the Neoliberal University," *Learning and Teaching: The International Journal of Higher Education in the Social Sciences* 7, no. 1 (Spring 2014): 12–28.
4. See Bawaka Country (including Sandie Suchet-Pearson, Sarah Wright, Kate Lloyd, and Laklak Burarrwanga), "Caring as Country: Towards an Ontology of Co-Becoming in Natural Resource Management," in "New Geographies of Coexistence: Reconsidering Cultural Interfaces in Resource and Environmental Governance," ed. Richard Howitt, Gaim James Lunkapis, Sandie Suchet-Pearson, and Fiona Miller, special issue, *Asia Pacific Viewpoint* 54, no. 2 (August 2013): 185–197; Robyn Dowling, "Geographies of Identity: Labouring in the 'Neoliberal' University," *Progress in Human Geography* 32, no. 6

(December 2008): 812–820.

5. Don Brenneis, Cris Shore, and Susan Wright, "Getting the Measure of Academia: Universities and the Politics of Accountability," *Anthropology in Action* 12, no. 1 (Spring 2005): 1–10; Dowling, "Geographies of Identity"; Margaret Jolly, "Antipodean Audits: Neoliberalism, Illiberal Governments, and Australian Universities," *Anthropology in Action* 12, no. 1 (Spring 2005): 31–47; Alison Mountz, Anne Bonds, Becky Mansfield, Jenna Loyd, Jennifer Hyndman, Margaret Walton-Roberts, Ranu Basu, Risa Whitson, Roberta Hawkins, Trina Hamilton, Winifred Curran, aka the Great Lakes Feminist Geography Collective, "For Slow Scholarship: A Feminist Politics of Resistance through Collective Action in the Neoliberal University," *ACME: An International E-Journal for Critical Geographies* 14, no. 4 (August 18, 2015), http://ojs.unbc.ca/index.php/acme/article/view/1058.
6. Burarrwanga and Family, *Welcome to My Country*, 25; Bawaka Country, "Caring as Country," 187.
7. Victoria Lawson, "Geographies of Care and Responsibility," *Annals of the Association of American Geographers* 97, no. 1 (2007): 1–11; Jeff Popke, "The Spaces of Being In-Common: Ethics and Social Geography," in *The SAGE Handbook of Social Geographies*, ed. Susan J. Smith, Rachel Pain, Sallie A. Marston, and John Paul Jones III (London: Sage, 2010); and Deborah Bird Rose, "Recursive Epistemologies and an Ethics of Attention," in *Extraordinary Anthropology: Transformations in the Field*, ed. Jean-Guy A. Goulet and Bruce G. Miller (Lincoln: University of Nebraska Press, 2007).
8. Sue Jackson and Louise Crabtree, "Politically Engaged Geographical Research with the Community Sector: Is It Encouraged by Australia's Higher Education and Research Institutions?," *Geographical Research* 52, no. 2 (May 2014): 146–156; Donna Haraway, *Modest_Witness@Second_Millennium.FemaleMan©_Meets_OncoMouse™: Feminism and Technoscience* (New York: Routledge, 1997); Nancy C. M. Hartsock, "The Feminist Standpoint: Developing the Ground for a Specifically Feminist Historical Materialism," in *Discovering Reality: Feminist Perspectives on Epistemology, Metaphysics, Methodology, and Philosophy of Science*, ed. Sandra Harding and Merrill B. Hintikka, 2nd ed. (Dordrecht, Netherlands: Kluwer, 2003), 283–310; Sarah Wright, "Knowing Scale: Intelle©tual Property Rights, Knowledge Spaces, and the Production of the Global," *Social & Cultural Geography* 6, no. 6 (December 2005): 903–921; David Turnbull, "Reframing Science and Other Local Knowledge Traditions," *Futures* 29, no. 6 (August 1997): 551–562.
9. However, they recently have been told by Laklak, Ritjilili, Merrkiyawuy, and Banbapuy that they now can proudly say they are toddlers!
10. Kate Lloyd, Sarah Wright, Sandra Suchet-Pearson, Laklak Burarrwanga, and Paul Hodge,

"L'entrelacement des vies: Les pratiques collaboratives sur le terrain dans le Nord-Est de la terre Arnhem, Australie" [Weaving lives together: Collaborative fieldwork in North East Arnhem Land, Australia], *Annales de Géographie* 212, no. 687/688 (September–December 2012): 513–524; Sandie Suchet-Pearson, Sarah Wright, Kate Lloyd, Laklak Burarrwanga, and Paul Hodge, "Footprints across the Beach: Beyond Researcher-Centered Methodologies," in *A Deeper Sense of Place: Stories and Journeys of Collaboration in Indigenous Research*, ed. Jay T. Johnson and Soren C. Larsen (Corvallis: Oregon State University Press, 2013), 29–32; Sarah Wright, Kate Lloyd, Sandie Suchet-Pearson, Laklak Burarrwanga, Matalena Tofa, and Bawaka Country, "Telling Stories in, through and with Country: Engaging with Indigenous and More-than-Human Methodologies at Bawaka, NE Australia," *Journal of Cultural Geography* 29, no. 1 (February 2012): 48; see also Julie Cupples and Sara Kindon, "Far from Being 'Home Alone': The Dynamics of Accompanied Fieldwork," *Singapore Journal of Tropical Geography* 24, no. 2 (July 2003): 211–228; Marie D. Price, "The Kindness of Strangers," *Geographical Review* 91, no. 1–2 (January–April 2001): 143–150; Paul F. Starrs, Carlin F. Starrs, Genoa I. Starrs, and Lynn Huntsinger, "Fieldwork . . . with Family," *Geographical Review* 91, no. 1–2 (January–April 2001): 74–87.

11. Sarah Amsler and Sara C. Motta, "The Marketised University and the Politics of Motherhood," *Gender and Education*, vol. 10 (2017), http://dx.doi.org/10.1080/09540253.2017.1296116; Motta, "The Messiness of Motherhood in the Marketized University," *Ceasefire*, June 14, 2012, https://ceasefiremagazine.co.uk/messiness-motherhood-marketised-university/.
12. Annelie Bränström-Öhman, "Leaks and Leftovers: Reflections on the Practice and Politics of Style in Feminist Academic Writing," in *Emergent Writing Methodologies in Feminist Studies*, ed. Mona Livholts (New York: Routledge, 2012); Mountz et al., "For Slow Scholarship"; Rachel Pain, "Impact: Striking a Blow or Walking Together?," *ACME: An International E-Journal for Critical Geographies* 13, no. 1 (2014), http://ojs.unbc.ca/index.php/acme/article/view/986/840; Noel Castree, "Research Assessment and the Production of Geographical Knowledge," *Progress in Human Geography* 30, no. 6 (2006): 749; Brenneis, Shore, and Wright, "Getting the Measure."
13. Jackson and Crabtree, "Politically Engaged Geographical Research," 153.
14. Richie Howitt and Stan Stevens, "Cross-Cultural Research: Ethics, Methods, and Relationships," in *Qualitative Research Methods in Human Geography*, ed. Iain Hay, 3rd ed. (Melbourne: Oxford University Press, 2010); J. K. Gibson-Graham, *The End of Capitalism (as We Knew It): A Feminist Critique of Political Economy* (Minneapolis: University of Minnesota Press, 2006); mrs kinpaisby, "Boundary Crossings: Taking Stock of Participatory Geographies: Envisioning the Communiversity," *Transactions of*

the Institute of British Geographers 33, no. 3 (July 2008): 292–299; Aileen M. Moreton-Robinson and Maggie Walter, "Indigenous Methodologies in Social Research," in *Social Research Methods: An Australian Perspective*, ed. Maggie Walter, 2nd ed. (Melbourne: Oxford University Press, 2009); Renee Pualani Louis, "Can You Hear Us Now? Voices from the Margin: Using Indigenous Methodologies in Geographic Research," *Geographical Research* 45, no. 2 (June 2007): 130–139; Linda Tuhiwai Smith, *Decolonizing Methodologies: Research and Indigenous Peoples*, 2nd ed. (London: Zed Books, 2012).

15. Mountz et al., "For Slow Scholarship," 8.
16. kinpaisby, "Boundary Crossings"; Deirdre Conlon, Nicholas Gill, Imogen Tyler, and Ceri Oeppen, "Impact as Odyssey," *ACME: An International E-Journal for Critical Geographies* 13, no. 1 (2014): 33–38; Brenneis, Shore, and Wright, "Getting the Measure"; Shore and Davidson, "Beyond Collusion"; Courtois and O'Keefe, "Precarity in the Ivory Cage," 45.
17. Being invited to contribute to this book highlighted how we had not discussed many of these issues in our work to date. Yet the range of personal difficulties we faced in writing up some of this material for this chapter—"Can we say this?" "Can we reveal that?" "How do we describe such and such?"—reveals these tricky spaces and murky zones and underscore just how problematic these issues are.
18. For discussion of different accountabilities in the context of natural resource management, see Samantha Muller, "Accountability Constructions, Contestations, and Implications: Insights from Working in a Yolŋu Cross-Cultural Institution, Australia," *Geography Compass* 2, no. 2 (February 2008): 33–38. Muller's work argues that the concept of accountability for the common good does not easily translate across cultures. Mainstream structures of accountability that portray themselves as being objective, verifiable, or "value free" clash with Indigenous-defined versions of accountability to Yolŋu *Rom*, or law, which are characterized by close kinship and family ties.
19. Richie Howitt and Jessica McLean, "Towards Closure? Coexistence, Remoteness, and Righteousness in Indigenous Policy in Australia," *Australian Geographer* 46, no. 2 (April 2015); Christopher Mayes and Jenny Kaldor, "Don't Be Surprised by Abbott's Comments About 'Life-Style Choices,'" *The Conversation*, March 12, 2015, accessed May 15, 2016, https://theconversation.com/dont-be-surprised-by-abbotts-comments-about-lifestyle-choices-38711; Shalailah Medhora, "Remote Communities Are 'Lifestyle Choices,' Says Tony Abbott," *The Guardian*, March 10, 2015, accessed May 15, 2016, http://www.theguardian.com/australia-news/2015/mar/10/remote-communities-are-lifestyle-choices-says-tony-abbott; Laklak Burarrwanga et al., Welcome to My Country.
20. Deborah Bird Rose, "Indigenous Ecologies and an Ethic of Connection," in *Global Ethics and Environment*, ed. Nicholas Low (London: Routledge, 1999). The Northern Territory

Intervention refers to a highly problematic and contested federal government policy since the late 2000s.

21. Nancy M. Williams, *The Yolŋu and Their Land: A System of Land Tenure and the Fight for Its Recognition* (Canberra: Australian Institute of Aboriginal Studies, 1986); Howitt and McLean, "Towards Closure."
22. Judith Halberstam, *The Queer Art of Failure* (Durham, NC: Duke University Press, 2011), 3.

BIBLIOGRAPHY

Amsler, Sarah, and Sara Motta. "The Marketized University and the Politics of Motherhood." *Gender and Education* 10 (2017). Http://dx.doi.org/10.1080/09540253.2017.1296116.

Bawaka Country (including Sandie Suchet-Pearson, Sarah Wright, Kate Lloyd, and Laklak Burarrwanga). "Caring as Country: Towards an Ontology of Co-Becoming in Natural Resource Management." In "New Geographies of Coexistence: Reconsidering Cultural Interfaces in Resource and Environmental Governance," edited by Richard Howitt, Gaim James Lunkapis, Sandie Suchet-Pearson, and Fiona Miller. Special issue, *Asia Pacific Viewpoint* 54, no. 2 (August 2013): 185–197.

Bawaka Country (including Sarah Wright, Sandie Suchet-Pearson, Kate Lloyd, Laklak Burarrwanga, Ritjilili Ganambarr, Merrkiyawuy Ganambarr-Stubbs, Banbapuy Ganambarr, Djawundil Maymuru, and Jill Sweeney). "Co-Becoming Bawaka: Towards a Relational Understanding of Place/Space." *Progress in Human Geography* 40 (August 2016): 455–475. Http://dx.doi.org/10.1177/0309132515589437.

Bränström-Öhman, Annelie. "Leaks and Leftovers: Reflections on the Practice and Politics of Style in Feminist Academic Writing." In *Emergent Writing Methodologies in Feminist Studies*, edited by Mona Livholts, 27–40. New York: Routledge, 2012.

Brenneis, Don, Cris Shore, and Susan Wright. "Getting the Measure of Academia: Universities and the Politics of Accountability." *Anthropology in Action* 12, no. 1 (Spring 2005): 1–10.

Burarrwanga, Laklak, Merrkiyawuy Ganambarr-Stubbs, Banbapuy Ganambarr, Sandie Suchet-Pearson, Kate Lloyd, and Sarah Wright. "Learning from Indigenous Conceptions of a Connected World." In *Enough for All Forever: A Handbook for Learning about Sustainability*, edited by Joy Murray, Glenn Cawthorne, Christopher Dey, and Chris Andrew, 3–13. Champaign, IL: Common Ground, 2012.

Burarrwanga, Laklak, and Family. *Welcome to My Country*. Melbourne: Allen & Unwin, 2013.

Castree, Noel. "Research Assessment and the Production of Geographical Knowledge." *Progress in Human Geography* 30, no. 6 (December 2006): 747–782.

Conlon, Deirdre, Nicholas Gill, Imogen Tyler, and Ceri Oeppen. "Impact as Odyssey." *ACME: An*

International E-Journal for Critical Geographies 13, no. 1 (2014): 33–38.

Courtois, Aline, and Theresa O'Keefe. "Precarity in the Ivory Cage: Neoliberalism and Casualisation of Work in the Irish Higher Education Sector." *Journal for Critical Educational Policy Studies* 13, no. 1 (June 2015): 43–66.

Cupples, Julie, and Sara Kindon. "Far from Being 'Home Alone': The Dynamics of Accompanied Fieldwork." *Singapore Journal of Tropical Geography* 24, no. 2 (July 2003): 211–228.

Dowling, Robyn. "Geographies of Identity: Labouring in the 'Neoliberal' University." *Progress in Human Geography* 32, no. 6 (December 2008): 812–820.

Gibson-Graham, J. K. *The End of Capitalism (as We Knew It): A Feminist Critique of Political Economy*. Minneapolis: University of Minnesota Press, 2006.

Halberstam, Judith. *The Queer Art of Failure*. Durham, NC: Duke University Press, 2011.

Haraway, Donna. *Modest_Witness@Second_Millennium.FemaleMan©_Meets_OncoMouse™: Feminism and Technoscience*. New York: Routledge, 1997.

Hartsock, Nancy C. M. "The Feminist Standpoint: Developing the Ground for a Specifically Feminist Historical Materialism." In *Discovering Reality: Feminist Perspectives on Epistemology, Metaphysics, Methodology, and Philosophy of Science*, edited by Sandra Harding and Merrill B. Hintikka, 2nd ed., 283–310. Dordrecht, Netherlands: Kluwer, 2003.

Howitt, Richie, and Jessica McLean. "Towards Closure? Coexistence, Remoteness, and Righteousness in Indigenous Policy in Australia." *Australian Geographer* 46, no. 2 (April 2015): 137–145.

Howitt, Richie, and Stan Stevens. "Cross-Cultural Research: Ethics, Methods, and Relationships." In *Qualitative Research Methods in Human Geography*, edited by Iain Hay, 30–50. 3rd ed. Melbourne: Oxford University Press, 2010.

Jackson, Sue, and Louise Crabtree. "Politically Engaged Geographical Research with the Community Sector: Is It Encouraged by Australia's Higher Education and Research Institutions?" *Geographical Research* 52, no. 2 (May 2014): 146–156.

Jolly, Margaret. "Antipodean Audits: Neoliberalism, Illiberal Governments, and Australian Universities." *Anthropology in Action* 12, no. 1 (Spring 2005): 31–47.

Lawson, Victoria. "Geographies of Care and Responsibility." *Annals of the Association of American Geographers* 97, no. 1 (2007): 1–11.

Lloyd, Kate, Sarah Wright, Sandra Suchet-Pearson, Laklak Burarrwanga, and Paul Hodge. "L'entrelacement des vies: Les pratiques collaboratives sur le terrain dans le Nord-Est de la terre Arnhem, Australie" [Weaving lives together: Collaborative fieldwork in North East Arnhem Land, Australia]. *Annales de Géographie* 212, no. 687/688 (September–December 2012): 513–524.

Louis, Renee Pualani. "Can You Hear Us Now? Voices from the Margin: Using Indigenous Methodologies in Geographic Research." *Geographical Research* 45, no. 2 (June 2007): 130–139.

Moreton-Robinson, Aileen M., and Maggie Walter. "Indigenous Methodologies in Social Research." In *Social Research Methods: An Australian Perspective*, edited by Maggie Walter, 1–18. 2nd ed. Melbourne: Oxford University Press, 2009.

Mountz, Alison, Anne Bonds, Becky Mansfield, Jenna Loyd, Jennifer Hyndman, Margaret Walton-Roberts, Ranu Basu, Risa Whitson, Roberta Hawkins, Trina Hamilton, and Winifred Curran, aka the Great Lakes Feminist Geography Collective. "For Slow Scholarship: A Feminist Politics of Resistance through Collective Action in the Neoliberal University." *ACME: An International E-Journal for Critical Geographies* 14, no. 4 (August 18, 2015). Http://ojs.unbc.ca/index.php/acme/article/view/1058.

mrs kinpaisby. "Boundary Crossings: Taking Stock of Participatory Geographies: Envisioning the Communiversity." *Transactions of the Institute of British Geographers* 33, no. 3 (July 2008): 292–299.

Muller, Samantha. "Accountability Constructions, Contestations, and Implications: Insights from Working in a Yolŋu Cross-Cultural Institution, Australia." *Geography Compass* 2, no. 2 (February 2008): 395–413.

Pain, Rachel. "Impact: Striking a Blow or Walking Together?" *ACME: An International E-Journal for Critical Geographies* 13, no. 1 (2014). Http://ojs.unbc.ca/index.php/acme/article/view/986/840.

Popke, Jeff. "The Spaces of Being In-Common: Ethics and Social Geography." In *The SAGE Handbook of Social Geographies*, edited by Susan J. Smith, Rachel Pain, Sallie A. Marston, and John Paul Jones III, 435–454. London: Sage, 2010.

Price, Marie D. "The Kindness of Strangers." *Geographical Review* 91, no. 1–2 (January–April 2001): 143–150.

Rose, Deborah Bird. "Indigenous Ecologies and an Ethic of Connection." In *Global Ethics and Environment*, edited by Nicholas Low, 175–187. London: Routledge, 1999.

———. *Nourishing Terrains: Australian Aboriginal Views of Landscape and Wilderness.* Canberra: Australia Heritage Commission, 1996. Https://www.environment.gov.au/system/files/resources/62db1069-b7ec-4d63-b9a9-991f4b931a60/files/nourishing-terrains.pdf.

———. "Recursive Epistemologies and an Ethics of Attention." In *Extraordinary Anthropology: Transformations in the Field*, edited by Jean-Guy A. Goulet and Bruce G. Miller, 88–102. Lincoln: University of Nebraska Press, 2007.

Shore, Cris, and Miri Davidson. "Beyond Collusion and Resistance: Academic-Management

Relations within the Neoliberal University." *Learning and Teaching* 7, no. 1 (Spring 2014): 12–28.

Smith, Linda Tuhiwai. *Decolonizing Methodologies: Research and Indigenous Peoples*. 2nd ed. London: Zed Books, 2012.

Starrs, Paul F., Carlin F. Starrs, Genoa I. Starrs, and Lynn Huntsinger. "Fieldwork . . . with Family." *Geographical Review* 91, no. 1–2 (January–April 2001): 74–87.

Suchet-Pearson, Sandie, Sarah Wright, Kate Lloyd, Laklak Burarrwanga, and Paul Hodge. "A Footprint in a Rock: Entwining Lives and Co-constructing 'the Field' in Australia." In *A Deeper Sense of Place: Stories and Journeys of Collaboration in Indigenous Research*, edited by Jay T. Johnson and Soren C. Larsen, 21–40. Corvallis: Oregon State University Press, 2013.

Turnbull, David. "Reframing Science and Other Local Knowledge Traditions." *Futures* 29, no. 6 (August 1997): 551–562.

Williams, Nancy M. *The Yolŋu and Their Land: A System of Land Tenure and the Fight for Its Recognition*. Canberra: Australian Institute of Aboriginal Studies, 1986.

Wright, Sarah. "Knowing Scale: Intelle©tual Property Rights, Knowledge Spaces, and the Production of the Global." *Social & Cultural Geography* 6, no. 6 (December 2005): 903–921.

Wright, Sarah, Kate Lloyd, Sandie Suchet-Pearson, Laklak Burarrwanga, Matalena Tofa, and Bawaka Country. "Telling Stories in, through and with Country: Engaging with Indigenous and More-than-Human Methodologies at Bawaka, NE Australia." *Journal of Cultural Geography* 29, no. 1 (February 2012): 39–60.

Ganawendamaw: Anishinaabe Concepts of Sustainability

Margaret Noodin

Apane igo aanikeshkodaadiyang ezhi-maadiziyang akiing.
[We are all endlessly replacing one another as we live on earth.]

Observe, connect, know yourself and every other self and all that is no self. Dream of ways to rekindle in darkness and move forward in the brilliant dawn. Visualize, celebrate, confront, grow, decompose and transform. Sustain yourself and others. *Oganawaabanjigewag, niganawaabandizo, giganawendamigoonaanig*. As a poet I know the difference between these words is the difference between fish in the sea, my own subconscious, and the earth's collective sustainability. As a linguist I know the literal translations: "They watch protectively," "I observe myself," and "We all look after all of you." That the use of endangered languages and literatures best fosters the teaching of endangered concepts of sustainable ecological relationships is the subject of this chapter. In isolation, these endangered languages and literatures each remain the "less commonly taught," the minority perspective, the postmodern, decolonial view, but when combined, they are the alloys of evolution.

In the book *Wisdom Sits in Places* (1996), Apache storyteller Dudley Patterson tells author Keith Basso: "Wisdom sits in places. It's like water that never dries up.

You need to drink water to stay alive, don't you? Well, you also need to drink from places. You must remember everything about them. You must learn their names. You must remember what happened at them long ago."[1] Throughout the world there is wisdom sitting in places waiting to be remembered. Many of the places in northeastern North America are remembered by the Anishinaabeg, a confederacy of communities that includes the Ojibwe, Odawa, and Potawatomi peoples. More than two hundred separate nations, located now in Canada and the United States, consider Anishinaabemowin their heritage language, and although each nation has a separate history, all are part of the Great Lakes Algonquian diaspora, with memories of water and land and life sustained through centuries of change.

The Great Lakes are ten-thousand-year-old children born on three-billion-year-old bedrock. The lakes are the largest single system of fresh surface water on earth, containing more than 18 percent of the world's supply.[2] These Sweetwater Seas are the center of the Anishinaabe universe. Creation stories begin with a coalescence of thought and matter followed by a connection between sky and sea with land growing like a seed, first into an island, then into all of the known Anishinaabe territory. Stories of floods allude to cataclysmic change and complex migration paths with many stopping places. This narrative history describes movement from east to west, but always, the people live near the lakeshore, the riverbed, the swamp, or the woods, connected by water below and above.

Stories of deluge and survival have been preserved in Anishinaabemowin. One of twenty-six Algonquian languages, Anishinaabemowin is no longer spoken as a first language in any home or community. Yet it lingers as an alternate worldview and aesthetic option in the hearts and minds of many. Estimates of the number of remaining speakers range from eight thousand to ten thousand, but all reports confirm that today the majority of fluent speakers who learned Anishinaabemowin as their first language are over seventy years in age. Like tides and temperatures, the number of voices fluctuates. Several immersion schools exist in Ontario, Wisconsin, and Minnesota, and more resources for learning are available each year, but the race against time is palpable as fluent elders take with them lifetimes of knowledge, and students struggle to begin their own journey to use Anishinaabemowin as a modern language. Much like the Great Lakes themselves, Anishinaabemowin has been threatened by modernity, industry, and invasive influences. The Endangered Species Act of 1973 tried to protect Kirtland's Warbler, pitcher thistle, lake iris, Karner's Blue Butterfly, blue pike, lake sturgeon, and other species native to the homelands of the Anishinaabeg in the Great

Lakes watershed. Two decades later, the Native American Languages Act of 1990 attempted to create pathways to revitalize more than two hundred indigenous languages in the United States. Critics would say society is nearly too late in both cases as global economics, climate change, educational disparities, and racism continue to get in the way of sustainable environmental, linguistic, and cultural diversity. Yet the time to appreciate linguistic and environmental systems is at hand, and when lessons are woven together, the lakes, the stories, and the cultures are sustained. In Anishinaabemowin and American Indian studies classes at the University of Wisconsin–Milwaukee, we work to see the connection between the water and the words along the shores of Lake Michigan.

The Voices Surrounding and Sustaining Us

The English word "sustainability" connotes a passive object that needs attention and is subject to an outside agency for its maintenance. In contrast, the Anishinaabe term *ganawendamaw* is a verb that connotes a spectrum of animacy for all life, allowing rocks, water, and humans to be described as coequal partners in the creation, maintenance, and evolution of a place. This basic concept conveys the idea of observation, protection, and preservation. Anishinaabemowin is a flexible language marked by constant word construction and variation. Teachers and storytellers have a variety of different words that could be considered roughly equivalent to the English term "sustainability." Despite this diversity of definitions, all fluent speakers recognize the core concept *ganawendamaw*. Marlene Stately, from the Leech Lake Band of Ojibwe in Minnesota, suggests *jiniganawejigandaagwak* as an active phrase explaining the way one might care for another. Alphonse Pitawanakwat, from Wikwemikong Unceded First Nation on Manitoulin Island in Ontario, uses *ganawendjigewin* and *naagadawendjigewin*, which he translates as "keeping track of sustenance." As speakers of Modern Anishinaabemowin, students at UWM learn language, culture, and environmental philosophy simultaneously as they work with the words to find them in primary texts or use them in new compositions.

Teaching sustainability in classes focused on the Anishinaabe language requires a broader view of society and science than most students initially bring to class. Anishinaabemowin is offered as an option for students to meet the university requirement of becoming proficient in a second language. Students expect to learn new words and grammar, but they are not usually prepared to stretch their worldview to include new paradigms of animacy and identity. As students study the stories of elders, the human view is defined as only one of many, including the voices of the wind, stones, trees, birds, and other beings. With long and careful observation, any part of the environment can have a "voice" and can begin to be viewed as part of a network of understanding and influence. Land, water, oil, and animals cease to be defined as resources subject to use and depletion dependent on the rights of citizens and nations. Instead, they are recognized as part of a complex system of life, death, and the space in between. They have a voice: not the anthropomorphized symbolic transfer of human values to nonhumans, but the genuinely other view of the world we all share. Geographer R. D. K. Herman has written, "In Indigenous sciences, the world is often understood in terms of flows of energies (and sometimes entities) across a permeable boundary between manifest and unmanifest realities."[3] Students who learn how to describe these realities observe the environment and reframe concepts of science and sustainability from an Anishinaabe perspective.

The first step is to observe *Anishinaabewakiing*. Although the literal translation of "*akiing*" is "land," the Anishinaabe concept of *Anishinaabewakiing* is closer to the scientific concept of "ecosystem," except it is an ecosystem that explicitly includes people, their culture, and history.[4] As we learn the words to describe all that surrounds and sustains us, I remind students in my classes to look for the connections. In complete agreement with the National Oceanic and Atmospheric Administration, we identify four primary land-habitat types within the broader Great Lakes ecosystem: *noopimikamig* (forests), *mashkode* (grassland), *mashkiig* (wetland), and *zaagawaamik* (dune) communities. Looking for linguistic clues to cultural perspectives, we also find the morpheme "*mash*" in *mashkode* and *mashkiig*, which is also found in the words *mashkiki* (medicine) and *mashkawizi* (strong). These words can lead to a discussion of strength and adaptation. If *mashkiki*, literally "strength from the earth," is the antidote for illness, how does this word influence the way speakers of Anishinaabemowin live with the earth?

Depending on the age and linguistic proficiency of the students, further connections can be discussed. For instance, how are the characteristics of the lake classified in Anishinaabemowin? As students hear the words *agwaayaashkaa* (tide comes in) and *animaashkaa* (tide goes out), they identify the *giiyaashkaa* (waves), which then can become *mamangishkaa* (huge waves). On the one hand, this is a simple exercise in vocabulary building. On the other hand, it is also critical inquiry in freshwater limnology that views all the components of the lake as animated and connected. Curiosity and empirical investigation reveal the difference between the verbs that express various relationships with the water. This foundation prepares students to understand relationships between such words as *jiikaa'ogo* (to have fun on the waves), *gonaba'ogo* (to be capsized), and *agwaa'o* (to be carried ashore or be drawn out of the water). Students of science and literature can ask: Which beings would experience these actions? Which beings could or could not control them? How are humans, fish, and algae different and alike? How is the water different *akwiindimaa* (where it is deep) and *dabasiindimaa* (where it is shallow)? How is the suffix "*aajiwan*" used to describe all possible characteristics of water in motion? How is the morpheme "*madaa*" used to trace liminal position in various contexts? For example:

- *madaagamin* (agitated water)
- *madaabii* (emerge from land toward the shore)
- *madaabiiyaabikaa* (to be on the rocky shore)
- *madaabiiyaakwaa* (to be on the wooded shore)

These are only a few of the words that begin with "*madaa*," and various rules allow speakers to create new words and modify others to describe the way beings interact with the shore. Patterns of word construction do not decode individual or communal philosophies, and many languages could speak of the same subjects. But the language used for centuries on these shores contains recurring patterns that have been artfully expanded to emphasize subjects considered significant by speakers and eventually storytellers who seek ways to understand and adapt to one place over time.

Reading works originally composed in Anishinaabemowin with light and literal translations allows alternate perspectives of biosystems to become apparent to students. More animate relationships offer equal opportunities to act, transform, learn, and teach. Anishinaabe people possess a fundamentally different view of the relationship between human beings and their surroundings, one that is based on a philosophy that simultaneously promotes the integrity of the environment and the well-being of those who reside there. Above all, that philosophy is based on the principle that the plants, animals, and minerals, which coexist with humankind, must be treated with the utmost respect.[5] The power and possibility of every entity is recognized in the strongest stories of sustainability.

Turning to the *Aadizookaanag*

Although there are two types of Anishinaabe stories, the stories discussed here are *aadizookaanag*. *Dibaajimowinan* are reports primarily of the present. The word *dibaajimowinan* is an inanimate noun derived from the verb *dibaajimo*, meaning "to recount." *Aadizookaanag* are stories of greater cultural significance as indicated by the fact that the word itself is an animate noun derived from the verb *aadizooke*, which means "to create a story that provides direction." In English, *aadizookaanag* are often spoken of as beings connected to all other life. In the valuable collection *Centering Anishinaabeg Studies: Understanding the World through Stories*, Kathleen Westcott explains that a traditional tale, an *aadizookan*, "participates in that river of life that flows into and through every being, in every moment of our lives."[6] Therefore, when looking for ways to understand how to sustain something alive that is changing, we turn to the *aadizookaanag*.

Central to the composition of *aadizookaanag* is the concept of motion, of journeying. The word for leading a good life, *minobimaadizi*, reflects this concept of moving forward in a good way. "*Mino*," the morpheme for good, modifies "*maadizi*," the verb for being alive. The little additional "*bi*" notes location and marks the fact that a life lived well has context and connects to a specific place. In Anishinaabe teachings on ecology, "land is known intimately through experience and journeying."[7] As elders tell stories in Anishinaabemowin, spatial and temporal relationships are revealed through description of biophysical elements and narration of ecological histories. Furthermore, the stories demonstrate principles of adaptive learning and social-ecological resilience. This observance and preservation of biophysical and cultural information is the narration of Anishinaabe sustainability. *Aadizookaanag* have long been referred to as "teaching tales," because they exemplify the process of education through experience and observation. In Anishinaabemowin, these characteristics are emphasized.

Another concept related directly to teachings about sustainability in traditional Anishinaabe stories is the role of elders and ancestors. Using the single Anishinaabe word *anikoobijigan* to equal the three English words "grandparent," "grandchild," and "ancestor," speakers are reminded the core verb is *ankoobjige* (to string together). In this example, it is easy to see the importance of continuity and preservation of the behaviors and beliefs transmitted from one generation to the next. For thousands of years, stories were transmitted orally and the language had no orthography. Although Anishinaabemowin has been written now for several centuries, most early documents focused on trade and religious conversion. Until the mid-1800s, the average Anishinaabe person was connected to the past and learned the details of ecology, botany, and biology through stories told by the elders in the community.

However, colonization and assimilation led to community use of writing and a willingness to archive Anishinaabe literature. In 1903 and 1905, Fox linguist William Jones spent time in the Great Lakes area recording stories of elders. In these tales we find glimpses of the way the lakes are an integral part of Anishinaabe identity and serve as repositories of social and scientific knowledge. Several of the stories he collected from various elders show how the Anishinaabeg recognize many forms of life in the biosphere, practice adaptive learning, and demonstrate social and ecological resilience.[8]

In the story "Wenabozho Steals Fire," told by Waasaagoneshkang, an Anishinaabe grandmother speaks of "*agaami-kichigami ayaawag igiw anishinaabeg* [the Anishinaabeg who live on the far shore of the sea]" and says "*geget aapiji oganawendaanaawaa i'iw isa endaawaad* [truly, a very careful watch do they keep over it there where they dwell]." Pointing out the way the others care for their home, she wants her grandson to learn the connection between protection and survival. Using the term *ganawendaan*, she asks her son to observe their ways; in fact, in this story, after he carefully observes the far-away Anishinaabeg, he sets out to "*gimoodid ishkode odayaanaawaa* [steal the fire they possess]." To reach their home, he altered the landscape by asking the water to freeze. "*Ambe sa noo da-gashkadin o'ow gichigami, wiigwaasabakwaang da-wii-apiitadin o'ow gichigami* [Now should the sea freeze, as thick as the birch-bark covering of the lodge this sea]." And then he *waaboozoonsiw*, literally "rabbited himself," which is to say he became like a rabbit in many essential ways while still also remaining himself. In this way, Wenabozho was able to bring fire to his grandmother. This story illustrates adaptive change and creative adaptation of the environment within the boundaries of the ecological system.

Sometimes the point of a parable is to demonstrate what should not be done. In another of the stories Jones transcribed, Midaasoganzh tells of "Wenabozho and the Dancing Bulrushes" to emphasize the importance of careful observation. The story begins, as many do, with a journey:

Ningoding igo babimosed awiya onoondawaa zaasaakwenid.

[Once, while traveling about, he heard the sound of someone whooping.]

Wegoneniiwinen onaazitawaa.

[He went to where he heard the sound.]

Goniginiin, zaaga'igaans zaagidawaanig i'iw ziibiins, mii imaa ayaanid i'iwe ininiwa', bizhishig i'iw ininiwa'.

[Where a pond flowed into a brook there was a group of men.]

Gakina bingwaashaagidiwa, gaye dash ezhi'onid gakina waabigoniin obadakibine'oni.

[They were all naked with only flowers sticking up on their heads.]

Zhigwa owaabamigoo.

[Eventually they saw him.]

"E! Wenabozho, niwii-niimi'idimin, nishwaasogon niwii-niimi'idimin."

["Hey! Wenabozho! We are going to dance for eight days."]

Not one to be left out, Wenabozho danced with them to the point of exhaustion and tears. Only after utter defeat did he realize as dawn broke on the eighth day that he had been dancing with *zhaashaaganashkoon* (the cattails.) Of course they called out to him, and of course he knew they were dancing; the only question is how he mistook these cattail people for Anishinaabeg. The explanation offered in the story is this: "*Mii nangwana iw apii ko i'iw gii-ani-dagwaagig mii nangwana i'iw gii-gichi-naanooding* [It happened to be autumn, the season when there was always a strong wind blowing]." The fact that Wenabozho was tricked so easily emphasizes the need to be attentive and not fooled by the elements. The story also blurs the lines between beings of the swamp, beings of the land, and beings who live in the sea.

In a third story from the same series, Wenabozho and his grandmother make a connection between peace and the environment. "When Wenabozho Was Swallowed by a Sturgeon," told by Gaagige-pinesi-kwe (Marie Syrette), begins with the existential ruminations of Wenabozho, who says to his grandmother, "*Indige sa mii go niin gaa-izhi-pezhigowaanen* [I am curious to know if I am the only one]." Her response sets him on an epic adventure too complex to recount here in full detail, but eventually concluding with Wenabozho and his canoe in the belly of a Great Lake sturgeon. Eventually, with the help of a flock of gulls, he is set free, and the gulls are gifted with their bright white color.

Throughout the story, numerous inhabitants of the freshwater sea are introduced, and the sea is a formidable character as well. Complex enough to rival the odyssey stories of any culture, the tales of Wenabozho teach the facts required for stewardship and partnership with the lake. This tale in particular returns several times to the subjects of war, peace, and the consequences of altering an environment. Grandmother reminds Wenabozho of the way the cycle of the sun and the day should produce contentment:

> *Bizaan igo ji-pimaadiziyan; enishigo waabandaman giizhig baaji-wayaaseyaagin; gaye zhayiigwa giizis ba-mooka'angin ezhi-onaanigwendaagwag, ji-izhi-onaanigwendaman.*
>
> [And that in peace you should live; that you should behold with a feeling of contentment the light of day when it comes; and that whenever the sun comes forth, when a sense of gladness pervades all things, you should be joyful too.]

Whether Wenabozho and his ancestors will find contentment in continual dawns is not known, but in the stories we find ideals that relate to cultural continuity.

As the Anishinaabe experienced significant erosion of their natural environment and social systems, the stories changed as well. The peace described by Wenabozho's grandmother became more elusive in the 1900s. In 1947 and 1948, a series of stories first told by Frank Ettawageshik Sr. were recorded by Jane Willet and appeared in a new translation in 2015 by Howard Webkamigad. In these stories, we find that fifty years has not changed the way the people maintain connections to the land, yet we begin to see how the old stories are faced with the challenges of modern life.

In these stories, people are still connected to places. For example, the setting for "The Anishinaabe Woman Who Married Her Brother-in-law" is a maple-sugaring camp, which allows the narrator to describe the details of the sugar bush and reinforce the fact that "*pane gonaa iwi ziisibaakod gonaa (doo) binakaazang, maanda gonaa (doo) wiisining gonaa (doo) dagowondjigaade maanda ziisibaakod* [the Anishinaabe always have used sugar; it was used when the people ate]." Furthermore, the story uses the familiar sustainability term, *ganawendan*, to tell us "*ganawendang ziisibaakodokaan* [the sugar camp was always watched and protected]."[9]

Webkamigad's translations of these Ottawa stories show that places are protected for their spiritual significance. For instance, Makwa Zhagishing (Sleeping Bear Dunes) is described as a site of individual meditation and learning:

> *Anishinaabek dash gonaa zhewe gii nigibeshiwok ko, gii gichi piitendaanaawaa iwi makwa zhagishing dash (doo) daming. gii manidookenaawaa gwonda anishinaabek zhaazhago iwi. gii gichi ishpidanaamigat iwi. miidash ko gaa nanakaadimowaat anishinaabek zhaazhago iwi, gii gagidaakiiyewaat zhewe, wii gikendamowaat giiyeN,*

> *mandj giiyeN iidik minik waa bimaadiziwaagoweN. miidash iwi, miidash go iwi gaa zhimashkowendamowaat giishpmaa giiyeN wiya (doo) giizhidaakiiyewaat zhewe, wii gaagiye bimaadizi giiyeN. mii gaa zhibagosendamowaat.*
>
> [The Anishinaabe people used to stop at this place, they even camped there, and they stayed there to wait out the storms, and also it was considered a sacred place. The Anishinaabe people used it as a place for ceremonies, such as the vision quest ritual. The dunes were very high; it was a high sandy hill. One of the things they used to do, was they would climb up the dunes, and those who made it up to the top would then fast up on top to get their vision, and learn how they were to live. One of the things they believed was that if one is able to make it to the top, that person will have a long life.]

The stories serve as cartographies of culture. They remain records of the ways the landscape is instrumental in sustaining the culture. But they also contain didactic elements warning of threats to the environment. Through a dream sequence in the story "How Corn Came to the Anishinaabe People," for example, a man is warned that the loss of traditional understanding could lead to cataclysmic environmental change:

> *Iwi go oonji gii aagonewitaw yin, nongom go maanda naaniibowi ying, iwi go maanda aki go dekokaadaami ying, maanda gaye go mitigowaaki, kina go maanda gdoo dibendaan nongom, gaawiin dash wiyii go gibeying. bekaanizit wii dibendaan maanda. gwii wonaachitoon.*
>
> [Because you doubted me, you will suffer the consequences, this land we are now standing on and the forests you see around us, these are yours right now as they were given to you by the Creator, but because you doubted me, you will not have this land for much longer to call your own. Someone else will occupy it as you will lose control over the land.][10]

This story is set in history at the point when corn was first introduced to the Great Lakes as a part of early agricultural tradition, a moment long before the manifestation of the current nation's destiny. As it was retold in the 1900s it also served as a warning against the dehumanization of industry and the material commodification of land. The story is botanically and politically accurate, connecting it solidly to the tradition of adaptive learning needed for cultural survival.

We Should All Remember Always

This chapter is only a brief introduction to a practice of linguistic and literary methodology that is part of my pedagogy. This type of close reading has long been done, of course, in other languages and literature classrooms. What this chapter attempts to highlight, however, is the continued need to create interdisciplinary sites of inquiry where endangered languages and endangered cultures can address the globally significant subject of sustainability. As a faculty member who crosses classroom, community, and linguistic borders, I am sometimes asked if I experience tensions as a university professor whose time, energies, and interests are dedicated toward community work. Of course I have encountered those who do not recognize the need to teach introductory Anishinaabemowin in order to build a program that leads to creative writing in Anishinaabemowin. And I have had to explain the value of linguistics to numerous students and fellow teachers in community settings. Yet I find this work has prepared me to be attentive, seek sustainability, and guard against imbalance. There is a time to celebrate spring thunders, a time to plant, a time to harvest, and a time to rest. There is a time to teach youth in communities. There is a time to produce creative work using Anishinaabemowin as a modern language. There is a time to archive linguistic data, and there is a time to teach and reflect with other minds on what is known.

Because I received tenure late in my career, I view its security as a gift rather than a right; a gift, according to the Anishinaabe stories I teach, is the mark of a relationship that requires work to maintain. I spent many years as a member of the precariat, but long ago I also used my liberal arts degree to launch a career in finance and management, which has served me well in higher education. In phonology, administration, wild-rice winnowing, and the construction of poems, every detail matters. Most important, to those who see in my work a way of remediating what it means to be a university educator, I suggest that it is so only because Anishinaabemowin and the study of Anishinaabe narratives have taught me to dance with the cattails for only one day and sleep in the dunes when I can. The only sign of success in education is the ability of your students' students.

Potawatomi scholar Robin Wall Kimmerer has written that "traditional knowledge is woven into and is inseparable from the social and spiritual context of the culture."[11] I was reminded of this recently as I stood beside a two-thousand-year-old burial mound, with colleagues from my campus, elders from my community, and families from the Milwaukee Indian Community School. We were educating our fellow citizens about the inheritance we share in southeastern Wisconsin, where thousands of effigy and burial mounds trace a past only partially remembered. These sites of interaction are endangered by property law and progress, and they need our voices to remind the current generation of their value. Sometimes it seems the only thing not endangered is endangerment, but where exactly would that thinking lead? Teaching Anishinaabemowin and sustainability in the face of great loss and cataclysmic change is not easy. The trajectory has not yet reversed for the language or the lakes. That day we sang a song I composed for the occasion:

Gimikwenimimigo giizis ezhi-zaagiiaasige
[We remember you sun how you shine]
Gimikwenimimigo dibiki-giizis baabaabii'o
[We remember you moon waiting]
Mikwendaagozidaa apane gosha
[We should all remember always]
Mikwendaagozidaa apane gosha
[We should all remember always]

Gimikwenimigom aanikoobijiganag
[We remember the connected ancestors]
Gimikwenimigom inawemaaganag
[We remember all our relatives]
Mikwendaagozidaa apane gosha
[We should all remember always]
Mikwendaagozidaa apane gosha
[We should all remember always][12]

Standing beside me in ten-degree weather with the shore of Lake Michigan in sight was the young coauthor of another song, which takes this chapter back to where it began. I have a habit of traveling to visit the Milwaukee Indian Community School and sharing hand-drum songs. Some days the group is large. One day, only two girls were there, Waasse Williams and Naeli Valerine. That day, I asked the girls if they might like to create their own song since there were just a few of us. They suggested that they'd like to create a new song that was better than a "bad song" they had heard.

Not knowing what these urban girls might have heard in this hyperconnected century, I asked if they wanted to share the old song so we could think about what might be better. They said it was just a no-good song that said, "Rain, rain go away; come again another day." These were girls of the city but also girls of the shoreline. Professors from UWM's School of Freshwater Sciences assure me the children of this school drink each day from an aquifer forty thousand years old. These children know of the wisdom in places. They had heard of water walks and moon ceremonies. They knew that disdain for water was not healthy, and they wanted a song that celebrated water. Of course we have a song to celebrate the water, but the simple song they created that day celebrated not just the water; it celebrated the connection between all lives supported by water. Quite powerfully, the song begins in Oneida, a language from the other side of the Great Lakes, and ends with Anishinaabemowin:

Ukwehuwe ni'i
[Oneida I am]
Loganolo
[It is raining]
Loganolese
[It is really raining]
Yoyanole ohnekanus
[It tastes good, the water]

Anishinaabe nd'aaw
[Anishinaabe I am]
Gimewon
[It is raining]
Chigimewon
[It is really raining]
Minogame nibi
[It tastes good, the water][13]

Aanii ezhi-gikinoo'amawangidwaa zaagichigaade zaaga'igan Anishnaabemong? How do we teach the Anishinaabe way of loving the lakes? Is it possible to rekindle our endangered languages and humanity's endangered relationship with Chigaming, our vast freshwater sea? My best answer so far is to sing. We sing as scholars and students. We sing after listening to elders who appear in our dreams. We sing to the trees, lyrics learned from the wind. We sing as we travel through each day, which matters only as much as every day before and every day that follows in a sustainable chain of knowing how to observe, connect, know your self and every other self and all that is no self.

NOTES

1. Keith H. Basso, *Wisdom Sits in Places: Landscape and Language among the Western Apache* (Albuquerque: University of New Mexico Press, 1996), 70.
2. Government of Canada, Environment Canada, Ontario Region, and U.S. Environmental Protection Agency, Great Lakes National Program Office, *The Great Lakes: An Environmental Atlas and Resource Book*, ed. Kent Fuller, Harvey Shear, and Jennifer Wittig, 3rd ed. (Chicago: Environment Canada and U.S. Environmental Protection Agency, 1995), 6.
3. R. D. K. Herman, "Reflections on the Importance of Indigenous Geography," *American Indian Culture and Research Journal* 32, no. 3 (2008): 75.
4. Fikret Berkes and Iain Davidson-Hunt, "Learning as You Journey: Anishinaabe Perception of Social-Ecological Environments and Adaptive Learning," *Ecology and Society* 8, no. 1 (2004): 20.
5. Leanne R. Simpson and Paul Driben, "From Expert to Acolyte: Learning to Understand the Environment from an Anishinaabe Point of View," *American Indian Culture and Research Journal* 24, no. 3 (2000): 14.
6. Margaret Noodin, *Bawaajimo: A Dialect of Dreams in Anishinaabe Language and Literature* (East Lansing: Michigan State University Press, 2014), 42; Eva Marie Garrote and Kathleen Delores Westcott, "The Story Is a Living Being: Companionship with Stories in Anishinaabeg Studies," in *Centering Anishinaabeg Studies: Understanding the World through Stories*, ed. Jill Doerfler, Niigaanwewidam James Sinclair, and Heidi Kiiwetinepinesiik Stark (East Lansing: Michigan State University Press, 2013), 68.
7. Fikret Berkes et al., "Exploring the Basic Ecological Unit: Ecosystem-Like Concepts in Traditional Societies," *Ecosystems* 1, no. 5 (September 1998): 412.
8. All quotations from the Wenabozho stories used in this essay are taken from the extraordinarily helpful site *Baadwewedamojig* ("those who come making a sound"), created in 2012 by Alan Corbiere of M'Chigeeng First Nation and Alana Johns of the University of Toronto. To further clarify matters, a single spelling of "Wenabozho" (also known as Nenabosho, Nenaboozhoo and Nanabushu) has been used. While archivists

and linguists may wish to explore the reasons for each variation, storytellers and audiences would have been aware of all the variations at the time, and they remain aware today of the mercurial nature of Wenabozho, the Anishinaabe teacher and trickster.

It is important to note that these stories carry with them a complex orthographic history. The original transcriptions by William Jones, edited by Truman Michelson, contain a range of diacritics and dashes used by linguists of the 1900s. This system of writing was an attempt to preserve the oral presentation of the stories, but none of the speakers at the time had the expertise needed to edit her or his own work. This system is no longer recognizable to modern readers of Anishinaabemowin today and was primarily a way for nonspeakers to represent what they heard.

9. Howard Webkamigad, ed. and trans., *Ottawa Stories from the Springs: Anishinaabe Dibaadjimowinan Wodi Gaa Binjibaamigak Wodi Mookodjiwong E Zhinikaadek* (East Lansing: Michigan State University Press, 2015), 265, 269.
10. Ibid., 371.
11. Robin Wall Kimmerer, "Weaving Traditional Ecological Knowledge into Biological Education: A Call to Action," *BioScience* 52, no. 5 (May 2002): 434.
12. Margaret Noodin, "Gimikwenimigo (We Remember You)," *The Language of the Three Fires Confederacy*, 2016, http://ojibwe.net/songs/womens-traditional/we-remember-you/.
13. Waasse Williams, Naeli Valerine, and Margaret Noodin, "Ohnekanus Nibi (Water)," *The Language of the Three Fires Confederacy*, 2013, http://ojibwe.net/songs/childrens-songs/ohnekanus-nibi-water/.

BIBLIOGRAPHY

Basso, Keith H. *Wisdom Sits in Places: Landscape and Language among the Western Apache.* Albuquerque: University of New Mexico Press, 1996.

Berkes, Fikret, and Iain Davidson-Hunt. "Learning as You Journey: Anishinaabe Perception of Social-Ecological Environments and Adaptive Learning." *Ecology and Society* 8, no. 1 (2004): 5–27.

Berkes, Fikret, Mina Kislalioglu, Carl Folke, and Madhav Gadgil. "Exploring the Basic Ecological Unit: Ecosystem-Like Concepts in Traditional Societies." *Ecosystems* 1, no. 5 (September 1998): 409–415.

Corbiere, Alan, and Alana Johns. *Baadwewedamojig.* Http://ojibweproject.weebly.com/.

Garrote, Eva Marie, and Kathleen Delores Westcott. "The Story Is a Living Being: Companionship with Stories in Anishinaabeg Studies." In *Centering Anishinaabeg Studies: Understanding the World through Stories*, edited by Jill Doerfler, Niigaanwewidam James

Sinclair, and Heidi Kiiwetinepinesiik Stark. East Lansing: Michigan State University Press, 2013.

Government of Canada, Environment Canada, Ontario Region, and U.S. Environmental Protection Agency, Great Lakes National Program Office. *The Great Lakes: An Environmental Atlas and Resource Book*. Edited by Kent Fuller, Harvey Shear, and Jennifer Wittig. 3rd ed. Chicago: Environment Canada and U.S. Environmental Protection Agency, 1995.

Herman, R. D. K. "Reflections on the Importance of Indigenous Geography." *American Indian Culture and Research Journal* 32, no. 3 (2008): 73–88.

Kimmerer, Robin Wall. "Weaving Traditional Ecological Knowledge into Biological Education: A Call to Action." *BioScience* 52, no. 5 (May 2002): 432–438.

Noodin, Margaret. *Bawaajimo: A Dialect of Dreams in Anishinaabe Language and Literature*. East Lansing: Michigan State University Press, 2014.

———. "*Gimikwenimigo* (We Remember You)." *The Language of the Three Fires Confederacy*. 2016. Http://ojibwe.net/songs/womens-traditional/we-remember-you/.

Jones, William, comp. *Ojibwa Texts*. Edited by Truman Michelson. Leyden, Holland: E. J. Brill Publishers, 1917.

Simpson, Leanne R., and Paul Driben. "From Expert to Acolyte: Learning to Understand the Environment from an Anishinaabe Point of View." *American Indian Culture and Research Journal* 24, no. 3 (2000): 1–19.

Webkamigad, Howard, ed. and trans. *Ottawa Stories from the Springs: Anishinaabe Dibaadjimowinan Wodi Gaa Binjibaamigak Wodi Mookodjiwong E Zhinikaadek*. East Lansing: Michigan State University Press, 2015.

Williams, Waasse, Naeli Valerine, and Margaret Noodin. "*Ohnekanus Nibi* (Water)." *The Language of the Three Fires Confederacy*. 2013. Http://ojibwe.net/songs/childrens-songs/ohnekanus-nibi-water/.

Epilogue

> Bless the poets, the workers for justice,
> the dancers of ceremony, the singers of heartache,
> the visionaries, all makers and carriers of fresh
> meaning—We will all make it through,
> despite politics and wars, despite failures
> and misunderstandings. There is only love.
>
> —Joy Harjo, "Dedication"

The years since 2006, when this collection's editors arrived at the University of North Carolina, Pembroke, have seen a steady stream of national and global environmental catastrophes: the Deepwater Horizon oil spill, the Fukushima Daichii nuclear disaster, Hurricanes Sandy and Matthew, the worst drought in recorded history in the western United States, the battles over the Keystone XL and the Dakota Access pipelines, plus the hottest years in recorded history, with 2015 "marking the largest margin by which an annual temperature record has been broken." One of those catastrophes, Hurricane Matthew, ripped through our region in 2016, flooded our campus and nearby Lumberton, displaced many of our students and their families, and closed K-12 schools for weeks—impacts we saw firsthand, impacts that found their way into our class discussions. More than one author in this collection has referenced the Center for Biological Diversity, which notes that "our planet is now in the midst of its sixth mass extinction of plants and animals—the sixth wave of extinctions in the past half-billion years. We're currently experiencing the worst spate of species die-offs since the loss of the dinosaurs 65 million years ago."[1] These circumstances, unlike in the past, have been caused almost entirely by human behaviors. Despite valiant, encouraging climate talks in Paris, the state of our environments—plant, animal, geological, aqueous, human—is bad and only getting worse. As if all this hasn't been enough, President Trump's executive orders rolling back climate regulations and choice of U.S. Environmental Protection Agency Director Scott Pruitt signal an unabated assault on the environment, thanks to a dismantled EPA and an emboldened and empowered fossil fuel industry.

Those of us in sustainability and higher education know and continue to face these facts, yet we do what we do so that our lives may "be a counter-friction to stop the machine" of environmental injustice, to use the words of Henry David Thoreau in "Civil Disobedience." By educating for sustainability within environments like those discussed in this collection, we can agree that "[we] do not lend [ourselves] to the wrong which [we] condemn." Still, separating ourselves from environmentally destructive practices has gotten much more complicated since Thoreau's day; retreating to a cabin in the woods cannot do the work we need to do. Many of the authors writing here must fill their tanks with gas more than once a week to commute to the universities at which they teach; adjuncts might travel between multiple institutions at sometimes great distances from each other. Others conduct research that requires international air travel, pressing down another heavy carbon footprint. These realities underscore the claim that "formal sustainability and environmental education is delivered by individuals isolated, under pressure and without the resources typical of other disciplines,"[2] a truth the authors here make visible as they strive, with varying degrees of success, to find livable solutions for themselves and the environment. Yet the authors in this collection share the belief that despite the ways we are hamstrung by our energy systems, despite the fact that even environmentally conscious choices often do environmental damage, we must remain dedicated to educating ourselves and our students about sustainable ways of living on earth, cultivating habits of critical thinking that yield conscious, empathetic, intelligent, and loving people who will engage in sustainable futures, whatever these may look like.

As the chapters in this collection make clear, none of us quite knows what our students' futures will look like, let alone our own or our planet's—an uncertainty exacerbated by the inability to create a sustainable balance of our personal and professional lives within specific ecological, political, economic, and social communities. So many of the chapters here resonate with the theme of mobility and identity, what it means to stay in a place—chosen or adopted—and what it means to leave, who we are when we make these choices, what we gain and what we lose, and the web of circumstances that leads us to make these choices in the first place. The editors of this collection were led to consider this phenomenon anew in 2013, when we learned that the author of the story we had been teaching in our environmental literature course, "Swamp Posse" (discussed in our earlier chapter, "By the Lumbee River with Chad Locklear's 'Swamp Posse'"), had returned to Pembroke and joined UNCP's staff. Why did Chad Locklear return to Pembroke, and what did his return mean to him in terms of Robeson County's sustainability as homeplace? Wanting to ask Locklear these questions led us to explore even more intriguing ones.

We had already known that Locklear was a UNCP alumnus who left for graduate school elsewhere, just as Vince, his mostly autobiographical protagonist in "Swamp Posse," had done. We contacted Locklear both to tell him how much we enjoyed teaching his story and to ask him if we might interview him about it. Locklear replied with enthusiasm to our request, and he surprised us by revealing that since the published version of "Swamp Posse" had appeared in *Pembroke Magazine* in 2006, he had "polished the story up a bit" and wanted "to share this revised version with [us]." Locklear described his more recent version of "Swamp Posse," which he had renamed "River Stories," as "essentially the same story with a few added sentences here and there and descriptions that support character development and some main points that I didn't get to add at the time." Locklear described the published version of his story, the one we had been teaching so enthusiastically for years, as "a rough draft" and felt "River Stories" was "the best version of the two and the one [he was] most pleased with."[3] Understandably, we were eager to read Locklear's revision.

The revisions Locklear made to "Swamp Posse" in "River Stories," we found, were subtle but significant in relation to the story's central themes: the tension around what it means to stay in one's place of origin or to leave and what it takes to sustain oneself in place *as* oneself. In a way, Locklear seemed to have recrafted not only the social environment of Robeson County, but in particular, Vince's perceptions of that environment. The Vince of "River Stories" becomes, in Locklear's revision, someone who has been out in a larger world and then returned, an editorial erasure of Vince's ambivalence about whether he could sustain himself permanently in his homeland. The Vince of "River Stories" shows no indication of wanting or planning to stay. Despite an appreciation of his heritage as a Lumbee with strong connections to the river and region, Vince's thoughts of the future are firmly directed toward his new life with his white, blonde girlfriend Sabrina in a larger world. Despite his graduate work on oral histories of Lumbee elders, Vince states early in "River Stories":

> I was already tired of being back here. I was grateful to my father for letting me live rent-free in one of his mobile homes temporarily, but I missed my downtown apartment and the city. Who would have thought that I would have become the city boy that I am today? And who would have thought that I would have ever left this place to begin with or would be working on my second master's degree or engaged to my beautiful Sabrina?

These sentences don't appear in the published "Swamp Posse." In "River Stories," these lines replace the sentence in "Swamp Posse" in which Vince thinks, "Honestly, I was already growing tired of my research, but the elders' voices must be heard for future generations."[4] The Vince of "River Stories" appears less interested in carrying forward to future generations the voices of Lumbee elders than in asserting his own voice in a life he chooses for himself with the woman he loves, bringing with him but moving beyond the environment of his origins.

Locklear's revised Vince also does not tell the character Seven the story of how the Lumbee people became mists in the swamps—a scene we discussed in our earlier chapter—to calm his friend when Seven becomes upset. We asked Locklear during an interview why he made some of these revisions to his story: How had he reconceived his characters, especially Vince, resulting in some of the marked differences between "Swamp Posse" and "River Stories"? About revising the scene in which Vince had originally told Seven the story of the Lumbee people resolving into mist to hide themselves in the swamps, Locklear told us that he felt that moment in "Swamp Posse" hadn't been "authentic." His choice to remove that storytelling moment in "River Stories" was because, he told us, "I feel like [Vince] should have been helpless instead of a hero in that scene." It strikes us that this is precisely the dilemma in which many of us in sustainability education—as faculty, grad students, parents, partners, activists, allies, and artists—find ourselves, within the matrices of our institutions while doing the work that matters to us: hoping to be heroes, but often recognizing the ways in which we remain helpless to effect change. Our challenge, as the authors here eloquently articulate through their various stories, is to continue finding ways to reinvigorate our personal and professional lives while defending the ecosystems in which we find ourselves. We can only hope to continue revising our stories consciously and thoughtfully, as Locklear has done with "Swamp Posse" and "River Stories," to craft more authentic lives for ourselves and our students, even as we care deeply for the places we are and have been.

Our efforts in sustainability education, when we think about them, at times seem puny given the enormity of the planetary challenge. We are grateful, then, for the projects that institutions of higher education have undertaken to become carbon-neutral: replacing light bulbs, installing solar panels, composting waste, and recycling. Yet such projects, while vital, cannot fully achieve long-term sustainability without the simultaneous implementation of visible, consistent, transdisciplinary curricula. While individuals by themselves might not have the capacity to turn things around, communities and collectives can and do. The individuals in this collection, on their own or as members of cohorts, demonstrate that sustainability is more than worth the attempt. Our work is worth the attempt now and also in the long run, as it influences other individuals—our students—whose similar work will ripple outward, like the pebble tossed in the pond, to create and sustain future collectives. Our graduates can and will build the human resources we need for the future—impacts we might never see for ourselves. The future is demanding, and all of us might vacillate between outrage, anguish, joy, and despair; sometimes we might feel like the narrator of Ralph Ellison's *Invisible Man* (1952), who asks, "Who knows but that, on the lower frequencies, I speak for you?"[5] Yet for those of us doing the work of educating for sustainability, we cannot accept that we function solely on the lower frequencies. Instead, we want to become—and the essays in this collection show that we *can* become—the frequency that has the power both to change higher education and to save the planet.

NOTES

1. National Oceanic and Atmospheric Administration, National Centers for Environmental Information, *State of the Climate: Global Analysis for Annual 2015*, January 2016, http://www.ncdc.noaa.gov/sotc/global/201513; Center for Biological Diversity, "The Extinction Crisis," Center for Biological Diversity, http://www.biologicaldiversity.org/programs/biodiversity/elements_of_biodiversity/extinction_crisis/.

2. Henry David Thoreau, "Civil Disobedience," *American Studies at the University of Virginia*, last modified September 1, 2009, http://xroads.virginia.edu/~hyper2/thoreau/civil.html; Stuart Pearson, Steven Honeywood, and Mitch O'Toole, "Not Yet Learning for Sustainability: The Challenge of Environmental Education in a University," *International Research in Geographical and Environmental Education* 14, no. 3 (2005): 175, http://dx.doi.org/10.1080/10382040508668349.
3. Chad Locklear, interview by authors, June 1, 2015; Chad Locklear, email message to authors, June 14, 2013; Chad Locklear, "River Stories," unpublished manuscript, 2013.
4. Locklear, "River Stories," 1; Chad Locklear, "Swamp Posse," *Pembroke Magazine* 38 (2006): 172.
5. Ralph Ellison, *Invisible Man* (New York: Vintage International, 1995), 581.

BIBLIOGRAPHY

Ellison, Ralph. *Invisible Man*. New York: Vintage International, 1995.

Harjo, Joy. *Conflict Resolution for Holy Beings*. New York: W.W. Norton, 2015.

Locklear, Chad. "Swamp Posse." *Pembroke Magazine* 38 (2006): 172–186.

Pearson, Stuart, Steven Honeywood, and Mitch O'Toole. "Not Yet Learning for Sustainability: The Challenge of Environmental Education in a University." *International Research in Geographical and Environmental Education* 14, no. 3 (2005): 173–186. Http://dx.doi.org/10.1080/10382040508668349.

Contributors

Bawaka Country, including Laklak Burarrwanga, Ritjilili Ganambarr, Merrkiyawuy Ganambarr-Stubbs, Banbapuy Ganambarr, Djawundil Maymuru, Kate Lloyd, Sarah Wright, Sandie Suchet-Pearson, and Paul Hodge, is an Indigenous and non-Indigenous, human-more-than-human research collective. Laklak, Ritjilili, Merrkiyawuy, and Banbapuy are four Indigenous sisters, elders, and caretakers for Bawaka Country in northeast Arnhem Land together with their daughter, Djawundil. Kate, Sarah, and Sandie are three non-Indigenous human geographers from Newcastle and Macquarie Universities who have been adopted into the family as granddaughter, sister, and daughter. Paul is a non-Indigenous human geographer who has worked with and supported the collaboration by contributing valuable insights into the theoretical literature and drawing on his experiences of taking students on-Country. Bawaka Country is the diverse land, water, human, and nonhuman animals (including the human authors of this paper), plants, rocks, thoughts, and songs that make up the Indigenous homeland of Bawaka in North East Arnhem Land, Australia. Ours is a story of lives entwined and of new places of being and belonging. It is also a collaborative narrative of unexpected transformations, embedded families, and the spirituality and agency of nonhuman elements in, of, and as the landscape. The group has worked together as a research collective since 2006 and has written two books and several academic and popular articles together.

Brianna R. Burke is an Assistant Professor of Environmental Humanities and American Indian Studies in the English Department at Iowa State University. Currently she is working on a book that explores the morphing ideology of what it means to be (considered) Animal in a world of decreasing resources. She publishes on environmental and social justice, as well as on the pedagogy of both.

Keely Byars-Nichols is Associate Professor of English and department chair at University of Mount Olive in Mount Olive, North Carolina. She is the author of *The Black Indian in American Literature* (2014). Her teaching and research interests include composition studies, multicultural studies, and critical food studies.

Jennifer L. Case's poetry and prose have appeared in journals such as *Orion*, *North American Review*, *ISLE: Interdisciplinary Studies in Literature & Environment*, *English Journal*, *Poet Lore*, and *Stone Canoe*, where her work received the 2014 Allen and Nirelle Galson Prize in Fiction. She is an assistant nonfiction editor of *Terrain.org* and teaches creative writing, environmental writing, and composition at the University of Central Arkansas.

Jesse Curran received her PhD in English from Stony Brook University, where she currently teaches courses in literature, writing, and the environmental humanities. Her poetry and essays have been published in numerous journals, including the *Emily Dickinson Journal*, the *Journal of Sustainability Education*, *Fourth River*, *The Hopper*, and *About Place*. She also holds certifications in hatha yoga and permaculture design, which practically inform her approach to sustainable education and community building.

Barbara George is a teaching fellow and doctoral candidate in the Literacy, Rhetoric, and Social Practice program at Kent State University. George's scholarly interests include ecocriticism, literacy ecologies, and the investigation of the role of language and communication in emerging theories of sustainability. She is particularly interested in contributing to scholarship surrounding issues of environmental justice.

Jane Haladay is Professor of American Indian Studies and a member of the Esther G. Maynor Honors College faculty at the University of North Carolina at Pembroke. She teaches courses in Native American and environmental literatures and introductory courses in American Indian Studies that incorporate service learning and writing enrichment. She holds a Ph.D. in Native American Studies with an emphasis in Feminist Theory and Research from the University of California, Davis, and an M.A. from the University of Arizona's American Indian Studies Program. She has published critical and creative work in *Orion Magazine, The Vassar Review, Wicazo Sa Review, American Indian Women of Proud Nations (Critical Indigenous and American Studies), Teacher-Scholar: The Journal of the State Comprehensive University, Mediating Indianness: Transatlantic Refractions, ISLE: Interdisciplinary Studies in Literature & Environment, American Indian Culture and Research Journal,* and *SAIL: Studies in American Indian Literatures,* among others. She has received awards for outstanding teaching and UNCP's Excellence in Service Learning Award.

Scott Hicks, Ph.D., is Professor of English and member of the Esther G. Maynor Honors College faculty at the University of North Carolina, Pembroke. A graduate of the University of North Carolina, Chapel Hill, and Vanderbilt University, he teaches classes in African American literature, environmental literature, first-year composition, and the humanities, and his research on African American and environmental literatures, teaching, and service-learning appear in *Arizona Quarterly*; *Callaloo*; *Environmental Humanities*; *A Greene Country Towne: Philadelphia, Ecology, and the Material Imagination*; *ISLE: Interdisciplinary Studies in Literature & Environment*; *"The Inside Light": New Critical Essays on Zora Neale Hurston*; *North Carolina Literary Review*; *Service Learning and Literary Studies in English*; *What Democracy Looks Like: A New Critical Realism for a Post-Seattle World*; and *Working on Earth: Class and Environmental Justice*. He has won awards at UNCP for excellence in teaching and service-learning, and he served as chair of UNCP's faculty senate from 2014 to 2016.

Richard House is faculty in English at Rose-Hulman Institute of Technology and a cofounder of the Home for Environmentally Responsible Engineering. House, a Full Professor, received a PhD from the University of California, Irvine. In addition to engineering communication and pedagogy, he has scholarly interests in sustainability and Shakespeare.

Mark Minster is faculty in English at Rose-Hulman Institute of Technology and a cofounder of the Home for Environmentally Responsible Engineering. Minster, an Associate Professor, received his PhD from Indiana University and teaches courses in the intersections of writing, literature, religion, and environment.

Margaret Noodin is currently Director of the Electa Quinney Institute for American Indian Education and Associate Professor of English at the University of Wisconsin-Milwaukee. She is the author of *Bawaajimo: A Dialect of Dreams in Anishinaabe Language and Literature* and *Weweni*, a collection of bilingual poems in Ojibwe and English. She also serves as editor of www.ojibwe.net.

Andrea Olive is an Assistant Professor of Political Science and Geography at the University of Toronto Mississauga. She has a PhD in political science from Purdue University and is the author of *Land, Stewardship, and Legitimacy: Endangered Species Policy in Canada and the United States* (2014) and *The Canadian Environment in Political Context* (2016).

Derek Owens is Professor of English and director of the Institute for Writing Studies at St. John's University in Queens, New York. He is the author of *Composition and Sustainability (Teaching for a Threatened Generation)* (2001), *Memory's Wake* (2011), and *Resisting Writings (and the Boundaries of Composition)* (1994).

Jennifer Schell is an Associate Professor of English at the University of Alaska Fairbanks. Her specialties include American literature, Arctic writing, print and visual culture, animal studies, and environmental humanities. Her book *"A Bold and Hardy Race of Men": The Lives and Literature of American Whalemen* was published in 2013. She recently completed a series of articles on the ecogothic, and she is currently working on two book manuscripts on extinction narratives.

Daniel Spoth is Assistant Professor of Literature at Eckerd College in St. Petersburg, Florida. He received his PhD in English from Vanderbilt University, and he has held teaching appointments there as well as at Hendrix College and the University of Central Arkansas. His primary research and teaching interests are Southern literature (particularly William Faulkner), modern American poetry (particularly Ezra Pound, William Carlos Williams, and Wallace Stevens), and the American novel.

His work has appeared in *ELH: English Literary History*, *Americana*, the *European Journal of American Culture*, *Journal of Ecocriticism*, *FORUM*, and other periodicals.

Corey Taylor is faculty in English at Rose-Hulman Institute of Technology and a cofounder of the Home for Environmentally Responsible Engineering. Taylor, an Associate Professor, received his PhD from the University of Delaware and teaches courses in writing, sustainability, and American literature. His research interests include sustainability studies, African American music, and American poetry.

Index